MCDP 1-3

# Tactics



U.S. Marine Corps

PCN 142 000002 00

DEPARTMENT OF THE NAVY
Headquarters United States Marine Corps
Washington, D.C. 20380-1775

30 July 1997

# FOREWORD

This publication is about winning in combat. Winning requires many things: excellence in techniques, an appreciation of the enemy, exemplary leadership, battlefield judgment, and focused combat power. Yet these factors by themselves do not ensure success in battle. Many armies, both winners and losers, have possessed many or all of these attributes. When we examine closely the differences between victor and vanquished, we draw one conclusion. Success went to the ar- mies whose leaders, senior and junior, could best focus their efforts—their skills and their resources—toward a decisive end. Their success arose not merely from excellence in techniques, procedures, and material but from their leaders' abilities to uniquely and effectively combine them. Winning in combat depends upon tactical leaders who can think creatively and act decisively.

This book pertains equally to all Marine leaders, whether their duties entail combat service support, combat support, or combat arms. It applies to the Marine air-ground task force commander as well as the squadron commander and the fire team leader. All Marines face tactical decisions in battle regardless of their roles. Tactical leaders must develop and hone their warfighting skills through study and practice. This publication serves as a guide for that professional development. It addresses the theory of tactics and its application in a chaotic and uncertain environment.

The concepts and ideas within this publication are battle-tested. Throughout our history, one of the most important reasons for the success of the United States Marine Corps has been the military skill of our leaders at every level of command. Through their tactical skill and battlefield judgment, our commanders achieved tactical and operational advantage at the decisive time and place.

This publication is a revision of Fleet Marine Force Manual 1-3, *Tactics*, of 1991 and supersedes it. Marine Corps Doctrinal Publication (MCDP) 1-3 fully retains the spirit, scope, and basic concepts of its predecessor. MCDP 1-3 further develops and refines some of those concepts; in particular, a new chapter has been added on exploiting success and finishing, and some of the original material has been reorganized and expanded.

*Tactics* is in consonance with MCDP 1, *Warfighting*, and the other Marine warfighting publications. Presuming an understanding of maneuver warfare, MCDP 1-3 applies it specifically to the tactical level of war. Like MCDP 1, it is not prescriptive but descriptive, providing guidance in the form of concepts and ideas. This publication establishes the Marine Corps' philosophy for waging and winning battles.

C. C. KRULAK
General, U.S. Marine Corps
Commandant of the Marine Corps

DISTRIBUTION: 142 000002 00

# Tactics

## Chapter 1

# Understanding Tactics

*"In tactics, the most important thing is not whether you go left or right, but why you go left or right."*[1]

—A. M. Gray

*"There is only one principle of war and that's this. Hit the other fellow, as quick as you can, and as hard as you can, where it hurts him the most, when he ain't looking."*[2]

—Sir William Slim

This book is about winning in combat. Winning requires a thorough understanding and knowledge of tactics. But what is tactics?

## An Art and a Science

Tactics is "the art and science of winning engagements and battles. It includes the use of firepower and maneuver, the integration of different arms and the immediate exploitation of success to defeat the enemy,"[3] as well as the sustainment of forces during combat. It also "includes the *technical* application of combat power, which consists of those techniques and procedures for accomplishing specific tasks *within* a tactical action."[4] This description is from Marine Corps doctrine and reflects our approach to tactics. What does it tell us?

Tactics refers to the concepts and methods we use to accomplish a particular objective in either combat or military operations other than war. In war, tactics is the application of combat power to defeat the enemy in engagements and battles. Combat power is the total destructive force we can bring to bear against the enemy; it is a unique product of a variety of physical, moral, and mental factors.[5] Tactics results in the actions and counteractions between opposing forces. It includes the use of maneuver, supported by the application and coordination of fires, to gain advantage in order to defeat the enemy.

In military operations other than war, tactics may be the schemes and methods by which we perform other missions, such as to control a crowd or to provide a secure environment for the delivery of food, medicine, or supplies to a nation or people in need.

As stated in the definition, tactics is a combination of art and science to gain victory over the enemy. The art of tactics lies in how we creatively form and apply military force in a given situation. It involves the creation, positioning, and maneuver of combat power. When do we flank the enemy, and when do we ambush him? When do we attack, and when do we infiltrate? How do we use speed and momentum to achieve a decisive advantage? This creativity is a developed capacity, acquired through education, practice, and experi- ence.

The science of tactics lies in the technical application of combat power. It includes mastering the techniques and procedures that contribute to the development of warfighting skills such as marksmanship, navigation, gunnery, and close air support. The execution of these techniques and procedures must become second nature for us; this requires intensive and continuous training. Without mastery of basic warfighting skills, artistry and creativity in their application are impossible.

Now that we have examined the art and science of tactics, let us look at how we use tactics to complement strategy and campaigning. Strategy and campaigning bring our forces to a particular place at a particular time. We use tactics to win in combat. A war typically involves many individual engagements that form a continuous fabric of activity. Sometimes a cluster of engagements flows together to make up a battle that may last for hours, days, or even several weeks. Tactical competence is indispensable to victory in such engagements and battles. Leaders at the operational and strategic levels use tactical victories to bring about success in the campaign and, ultimately, in the war as a whole.

In combat, our objective is victory. Sometimes this involves the complete destruction of the enemy's forces; at other times achieving victory may be possible by attacking the enemy's will to fight. The Marine Corps must be equally prepared to win during both situations—those in which the enemy forces must be completely destroyed (as during World War II), and those in which the complete destruction of the enemy's forces may not be necessary or even desirable. As the Commanding General of the 1st Marine Division in Des-ert Storm, stated, "Our focus was not on destroying everything. Our focus was on the Iraqi mind and getting behind [it]."[6] He knew that the path to victory did not lie in the total destruction of the Iraqi forces, but in undermining their will to fight.

## The Environment

The tactical arena is a dynamic, ever-changing environment. The complexity of this environment makes combat chaotic and unpredictable. As an example of confusion and chaos on the battlefield, consider the amphibious assault on the island of Tarawa in November 1943.

During the assault, the combination of high casualties, lack of effective communications, and disruption of the 2d and 8th Marine Regiments' landings on the assault beaches led to a chaotic and nearly disastrous situation for the 2d Marine Division. Units were decimated under heavy fire. Surviving Marines huddled together under a coconut log sea wall in intermingled units without effective communications. Landing craft carrying reinforcements and supplies could not make it over a coral reef to the landing beaches. Only through daring leadership, initiative, and teamwork were Marines able to get off the beach and annihilate the defending Japanese force.[7]

The violence of combat only increases the level of confusion and chaos. Robert Sherrod, a *Time* and *Life* correspondent at Iwo Jima, gave testimony to this chaos in what he called "war at its worst":

> The first night on Iwo Jima can only be described as a nightmare in hell. . . . About the beach in the morning lay the

> dead. They had died with the greatest possible violence. Nowhere in the Pacific have I seen such badly mangled bodies. Many were cut squarely in half. Legs and arms lay fifty feet from any body.[8]

Battle is the collision of opposing forces—animate, interactive, and unpredictable in behavior. Performance varies from week to week, day to day, and even hour to hour as a unit interacts with its environment and the enemy.

Military forces are complex systems consisting of individuals and equipment. They interact internally and externally in seemingly chaotic ways. As Clausewitz wrote, "A battalion is made up of individuals, the least important of whom may chance to delay things or . . . make them go wrong."[9] As Marines, we believe the actions of single individuals can have great impact in combat and can also make things go right. For example, Sergeant John Basilone as a machine gunner at Guadalcanal contributed "in large measure to the virtual annihilation of a Japanese regiment."[10] He steadfastly manned his position in the face of repeated wave-type assaults and was instrumental in breaking the enemy's ability to press the attack, forcing them to retreat without achieving their goals.

Battle is also influenced by a variety of external conditions—directions and missions established by authorities, terrain, weather, attitudes of the civilian populace—that often cannot be foreseen. The outcome of combat can only be anticipated in terms of probabilities.

Technology also affects the tactical environment—but not always as anticipated. Technology may reduce uncertainty, and it also may increase it. The Spartans, organized into phalanxes, attacked in close formation, making it easy to see and control one's forces. Today, tactical formations are less well-defined as distances between elements have increased, complicating command and control. Increased weapons lethality, communications range, and tactical mobility cause us to disperse forces over greater distances. War is more fluid as a result of technology. While the machine gun bogged down warfare in World War I, tactical innovations like the tank, the airplane, and the aircraft carrier made warfare more rapid and free-flowing in World War II.

Future battle is likely to become even more chaotic. Although combat in Operation Desert Storm was between fairly well-defined forces in a well-defined space, the forces and operating areas in Vietnam, Somalia, and Grenada were far less well-defined. Enemy units were dispersed and often hidden within the civilian population, making them hard to detect and harder to target. They converged at a time and place of their choosing. Future opponents may choose to fight in this manner to offset our overwhelming superiority in fire- power.

This chaotic environment also brings opportunity. Clausewitz wrote about combat, "No other human activity is so continuously . . . bound up with chance."[11] The challenge is to recognize opportunity when it occurs in the midst of chaos and uncertainty and to seize it to obtain a clear, unambiguous victory. When viewed through time, even the most chaotic of systems may reveal recurring patterns that may then be exploited. The experienced tactician will look for these recurring patterns that can be exploited to advantage.

## How we view combat and how we fight

How we view the combat environment in large part determines how we operate in it. There are two competing views of combat. Some see it in simple terms as if the battle and the environment represent a closed mechanical system. This "deterministic" view argues that combat is predictable. Among the advocates of this view are military theorists who seek prescriptive rules for battle and analysts who predict battle outcomes based upon force ratios. The other view is that combat is chaotic and uncertain. In this "probabilistic" view, battle is seen as a complex phenomenon in which participants interact with one another and respond and adapt to their environment. The probabilistic viewpoint sees combat as unpredictable. The distinctions between these two views of combat are im-portant. They drive the choices commanders make in combat.

The deterministic view of combat often leads to centralized control. It can be a recipe for micromanagement stifling the initiative subordinates need to deal with combat's inevitable uncertainties. Overly prescriptive orders and plans inhibit a unit's ability to cope with uncertainty and change. Eventually, the unit, inflexible and unable to adapt, may be overwhelmed by events.

The probabilistic view of combat recognizes that the complexity and uncertainty of war leads to a more decentralized approach to control. We place greater trust in subordinates to achieve a desired result. Through use of mission orders and commander's intent, subordinates are able to handle unforeseen situations and exploit opportunities that arise.

Marine Corps tactics are based on the probabilistic view of combat. We must be able to cope with uncertainty and operate in an ever-changing combat environment. We must be flexible and responsive to changes in the situation. There are no fixed rules that can be applied automatically, and every situation is different. As one tactics manual put it more than half a century ago: "The leader who frantically strives to remember what someone else did in some slightly similar situation has already set his feet on a well-traveled road to ruin."[12]

Leaders must remember that there are no fixed rules and no precise checklists, but there are bounds. That is why successful leaders study, train, and exercise their minds to improve tactical proficiency. We study examples of successes and failures not to emulate someone else's scheme, but to increase our own tactical understanding and competence.

## Marine Corps Tactics

The successful execution of Marine Corps tactics hinges on the thoughtful application of a number of tactical concepts so as to achieve success on the battlefield. Key among these concepts are *achieving a decision, gaining advantage, being faster, adapting, cooperating,* and *exploiting success*. Each of these concepts is discussed in detail later in this publication. Creative and practical employment of these ideas throughout the planning and execution of tactics leads to success. These concepts are not stand-alone ideas but are to be combined so as to achieve an effect that is greater than their separate sum. Part of the art and science of tactics lies in knowing where and when to apply these concepts and which combinations to use to achieve the desired effect.

The number and definition of these concepts are not fixed, and their order of presentation does not indicate their value. Marines may find in their studies new or slightly different ideas that may be just as important. These ideas are presented in this publication so that readers will think about how to achieve success on the battlefield. These concepts help to provide a framework for developing a tactical mindset that has long been a hallmark of Marine leaders, from corporal through general.

## Conclusion

Tactical excellence is the hallmark of a Marine Corps leader. We fight and win in combat through our mastery of both the art and the science of tactics. The art of tactics involves the creative and innovative use of maneuver warfare concepts, while the science of tactics requires skill in basic warfighting techniques and procedures. It is our responsibility as Marine leaders to work continuously to develop our own tactical proficiency and that of our Marines. Understanding the concepts presented in this publication provides a foundation for that development.

# Chapter 2

# Achieving a Decision

*"It follows, then, that the leader who would become a competent tactician must first close his mind to the alluring formulae that well-meaning people offer in the name of victory. To master his difficult art he must learn to cut to the heart of a situation, recognize its decisive elements and base his course of action on these."*[1]

—*Infantry in Battle*

*"We must be ruthlessly opportunistic, actively seeking out signs of weakness, against which we will direct all available combat power. And when the* decisive *opportunity arrives, we must exploit it fully and aggressively, committing every ounce of combat power we can muster and pushing ourselves to the limits of exhaustion."*[2]

—FMFM 1, *Warfighting*

Tactics is the employment of units in combat. The objective of tactics is to achieve military success through a decision in battle. Using tactical actions to achieve a decision is central to Marine Corps tactics.

In the past, military forces have often won only incremental gains when they sought victory—taking a hill here or a town there, pushing the front forward a few kilometers, or adding to the body count. Sometimes these incremental gains were the result of a competent enemy or the chaotic nature of war. Many times, however, commanders sought incremental gains as a means to achieve victory. This incrementalist view sees war as a slow, cumulative process and is best exemplified by the grinding attrition tactics seen on the Western Front in World War I. There the opponents were more or less evenly matched, and their tactics resulted in indecisive action. In Vietnam, where the opposing forces were quite dissimilar in their military capabilities, the incremental approach led to the U.S.'s overreliance on firepower and body counts. This, in turn, led to the conduct of military operations that were often irrelevant to the outcome of the war, even though a comparison of casualty ratios appeared favorable.

Therefore, the Marine Corps has embraced a more flexible, imaginative, and effective way to wage war: maneuver warfare. Marine success with this approach has been demonstrated in places like Grenada and the Persian Gulf. In contrast to tactics based on incremental attrition, tactics in maneuver warfare always aims at decisive action.

This does not mean, however, that combat should be viewed as a bloodless ballet of movement. Combat, especially at the tactical level of war, will be characterized by tough, brutal, and desperate engagements. We must remember that war is a violent clash of two opposing wills in which each side is trying to wrest advantage from the other. Our future enemies may not allow us to gain, maintain, or employ technological or numerical superiority. The future battle may be bloody and tough, and that makes it vitally important that Marine leaders strive to develop tactical proficiency.

What do we mean by achieving a decision? Take a moment to compare these two historical examples.

## ANZIO: A MODEL OF TACTICAL INDECISIVENESS

In late 1943, the Allies were searching for a way to alleviate the stalemate in Italy. The campaign had stalled around the Cassino front and resembled the trench warfare of World War I. In order to keep the pressure on the Germans, bypass the stubborn German defenses at Cassino, and capture Rome, a bold operation was envisioned. The U.S. Army's 3d Division and the British Army's 1st Division would make an amphibious landing at Anzio, about 35 miles south of Rome. (See figure.)

**Indecisiveness:** Anzio, 1944. The Opportunity

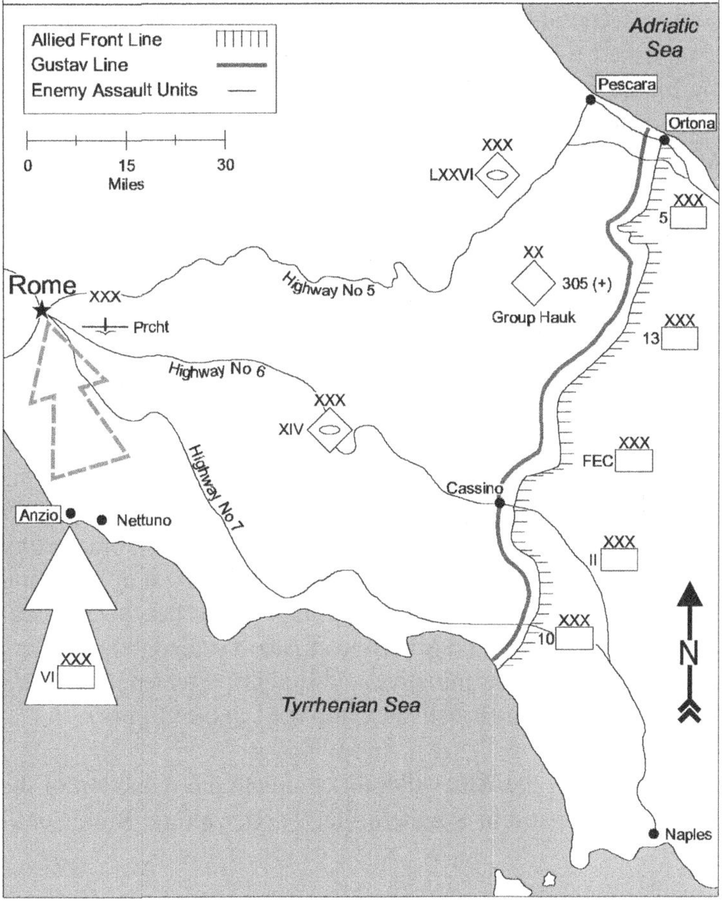

The Allies achieved complete surprise by landing at Anzio on January 22, 1944. Under the command of the U.S. Army's Major General Lucas, the Americans and British quickly established a beachhead and rapidly advanced 3 miles inland by midmorning against light German resistance. With the majority of their forces concentrated farther south around Cassino, the Germans could not possibly reinforce the Anzio beachhead until January 23d or 24th. If the Allies pressed their advantage, the road to Rome lay virtually undefended. The seizure of Rome would have had the effect of isolating the German defenders in the south and firmly establishing Allied control over Italy.

Yet General Lucas delayed. Concerned about being overextended and wanting to build up his logistics ashore, Lucas failed to press his initial advantage of surprise and allowed the Germans to reinforce the Anzio area. Not until January 29th did Lucas feel strong enough to make an offensive bid, but by that time it was too late. The Germans had arrived in force and had seized the dominating high ground in the beachhead area. Not only was the Allied offensive at Anzio stalled, but the Germans had seized the initiative and quickly threatened to drive the Americans and British back into the sea (see figure).

As a result, the Allies did not complete the reduction of the German defenses in southern Italy and capture Rome until

**Indecisiveness:** Anzio, 1944. Opportunity Lost

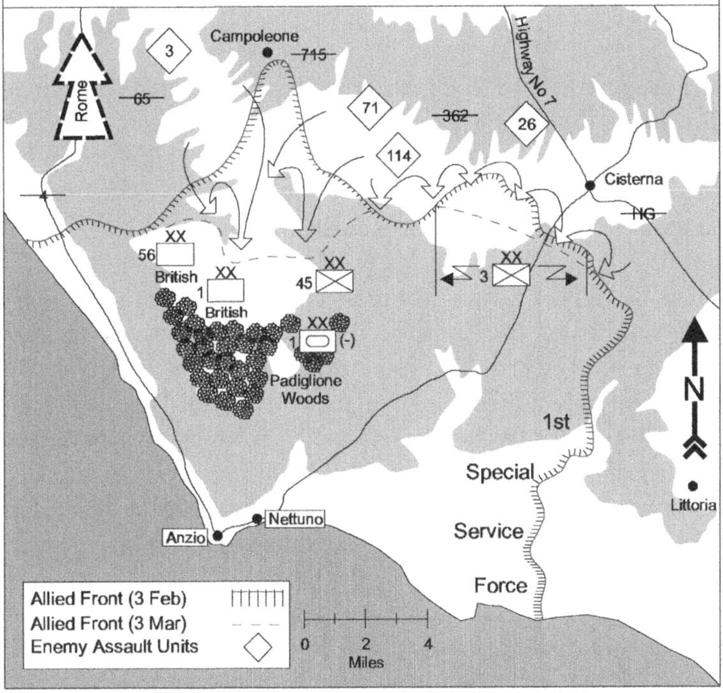


several months later. General Lucas lost a tremendous opportunity to exploit an initial success and gain a decisive result.[3]

## Cannae: A Clear Tactical Decision Achieved

On August 2, 216 B.C., the Carthaginian general Hannibal fought the Roman army under the command of Terentius Varro near the city of Cannae in southern Italy. Hannibal based his tactics on the specific characteristics of both forces and on the aggressive personality of the Roman commander.

As dawn broke, Hannibal drew up his force of 50,000 veterans with his left flank anchored on the Aufidus river, secured from envelopment by the more numerous Romans. His center contained only a thin line of infantry. His main force was concentrated on the flanks. His left and right wings each contained deep phalanxes of heavy infantry. Eight thousand cavalry tied the left of his line to the river. Two thousand cavalry protected his open right flank. Eight thousand men guarded his camp in the rear.

Varro and more than 80,000 Romans accepted the challenge. Seeing the well-protected Carthaginian flanks, Varro dismissed any attempt to envelop. He decided to crush his opponent by sheer weight of numbers. He placed 65,000 men in his center; 2,400 cavalry on his right; and 4,800 cavalry on his left and sent 11,000 men to attack the Carthaginian camp.

Following preliminary skirmishes, Hannibal moved his light center line forward into a salient against the Roman center. (See A in figure.) Then, his heavy cavalry on the left crushed

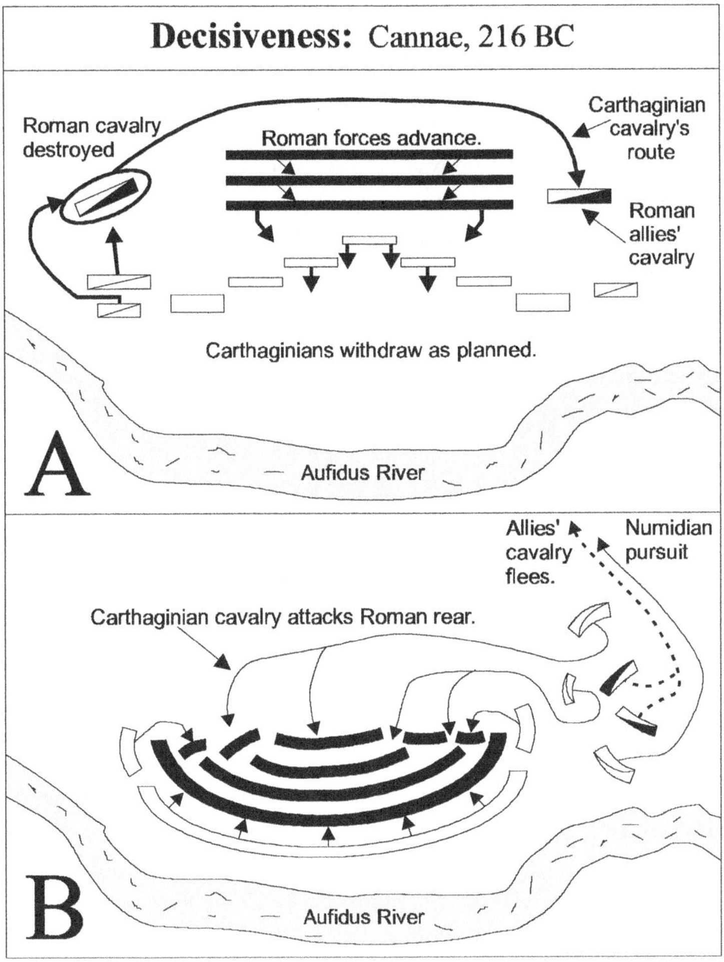
Decisiveness: Cannae, 216 BC
Roman cavalry destroyed
Roman forces advance.
Carthaginian cavalry's route
Roman allies' cavalry
Carthaginians withdraw as planned.
A
Aufidus River
Allies' cavalry flees.
Numidian pursuit
Carthaginian cavalry attacks Roman rear.
B
Aufidus River

the opposing Roman cavalry and swung completely around the Roman rear to attack the Roman cavalry on the other flank. The Roman cavalry fled the field.

The Carthaginian heavy cavalry then turned back to assault the rear of the dense Roman infantry who had pressed back Hannibal's thin center line. At the same time, Hannibal wheeled his right and left wings into the flanks of the Roman center. The Romans were boxed in, unable to maneuver or use their weapons effectively. (See B in figure on page 21.) Between 50,000 and 60,000 Romans died that day as Varro's army was destroyed.

## Understanding Decisiveness

What do these examples tell us about achieving a decision?

First, achieving a decision is *important.* An indecisive battle wastes the lives of those who fight and die in it. It wastes the efforts of those who survive as well. All the costs—the deaths, the wounds, the sweat and effort, the equipment destroyed or used up, the supplies expended—are suffered for little gain. Such battles have no meaning except for the comparative losses and perhaps an incremental gain for one side or the other.

Second, achieving a decision is *not easy*. History is litter- ed with indecisive battles. Sometimes it was enemy skill and determination that prevented even a victorious commander from achieving the decision he sought. In other cases, commanders fought a battle without envisioning a larger result for their actions. Sometimes, even with a vision of making the battle decisive, they could not achieve their goals due to the chaos and friction that is the nature of war and makes decisive victory so difficult.

That leads to the third lesson our examples point out. To be decisive, a battle or an engagement *must lead to a result beyond itself*. Within a battle, an action that is decisive must lead directly to winning in the campaign or war as a whole. *For the battle to be decisive, it must lead directly to a larger success in the war as a whole.*

On the other hand, we must not seek decisiveness for its own sake. We do not, after all, seek a decision if it is likely to be against us. We seek to ensure—insofar as this is possible, given the inherent uncertainties of war—that the battle will go our way. We have stacked the deck in our favor before the cards are laid on the table. Otherwise, to seek decisive battle is an irresponsible gamble.

When we seek battle, we must seek victory: accomplishment of the assigned mission that leads to further significant gains for the force as a whole. At Anzio, the Allied aim was to break the stalemate in the south, opening up a southern front that would force Germany to move additional forces from the defense of Normandy. This weakening of the Normandy defenses would support our planned invasion of France later that same year. At Cannae, Carthage won one round in its long contention with Rome for the domination of the Mediterranean. These tactical battles were planned for their overall operational and strategic effect. The consequences of a tactical engagement should lead to achieving operational and strategic goals.

## Military Judgment

Once we understand what is meant by the term decisive and why it is important to seek a decision, a question naturally arises: How do we do it?

There is no easy answer to that question; each battle will have its own unique answers. As with so much in warfare, it depends on the situation. No formula, process, acronym, or buzzword can provide the answer. Rather, the answer is in military judgment, in the ability of the commander to understand the battlefield and act decisively. Military judgment is a developed skill that is honed by the wisdom gained through experience. Combined with situational awareness, military judgment allows us to identify emerging patterns, discern critical vulnerabilities, and concentrate combat power.

**Understanding the Situation**

The first requirement of a commander is to understand the situation. The successful tactician studies the situation to develop in his mind a clear picture of what is happening, how it got that way, and how it might further develop. Consid- ering the factors of mission, enemy, terrain and weather, troops and support available-time available (METT-T), the commander must think through all actions, determine the desired result, and ascertain the means to achieve that result. Part of the commander's thinking should also include assuming the role of the enemy, considering what the enemy's best course of action may be, and deciding how to defeat it. Thinking through these elements helps the commander develop increased situational awareness.

Based on this understanding of the situation, the commander can begin to form a mental image of how the battle might be fought. Central to the commander's thinking must be the question, "In this situation, what efforts will be decisive?" The commander asks this question not just once, but repeatedly as the battle progresses. The commander must also address possible outcomes and the new situations that will result from those possibilities. As the situation changes, so will the solution and the actions that derive from it.

For every situation, the leader must decide which of the countless and often confusing pieces of information are important and reliable. The leader must determine what the enemy is trying to do and how to counter his efforts. The leader's skill is essentially one of *pattern recognition*, the ability, after seeing only a few pieces of the puzzle, to fill in the rest of the picture correctly. Pattern recognition is the ability to understand the true significance and dynamics of a situation with limited information. Pattern recognition is a key skill for success on the battlefield.

Tactics requires leaders to make decisions. A leader must make decisions in a constantly changing environment of friction, uncertainty, and danger. Making effective decisions and acting on those decisions faster than the enemy is a crucial element of Marine Corps tactics.

Sometimes there may be time to analyze situations deliberately and to consider multiple options. Comparing several options and selecting the best one is known as analytical decisionmaking. When time allows a commander to apply analytical decisionmaking—usually before an engagement or battle begins—the commander should make the most of it.

Once engaged, however, the commander finds time is short and the need for speed paramount. In some cases, speeding up the analytical decisionmaking process may be sufficient; however, in most cases intuitive decisionmaking is needed to generate and maintain tempo. Intuitive decisionmaking relies on a commander's intuitive ability to recognize the key elements of a particular problem and arrive at the proper decision without having to compare multiple options. Intuition is not some mysterious quality. Rather, it is a developed skill, firmly grounded in experience, and one that can be further developed through education and practice. It is not without some risk, however, and leaders should use the decisionmaking style that works for them.

Leaders with strong situational awareness and broad experience can act quickly because they have an intuitive understanding of the situation, know what needs to be done, and know what can be done. This insight has often been called *coup d'oeil* (pronounced koo dwee), a French term meaning literally "stroke of the eye." It has also been called "tactical sense."

Union Army Brigadier General John Buford's approach to the battle of Gettysburg offers a good example of understanding the battle so that it leads to a decision. Arriving at Gettysburg with a division of cavalry on the morning of June 30, 1863, Buford saw Confederate forces approaching from the northwest. With the bulk of the Union forces still some miles away, Buford was able to conceptualize the coming battle in his mind. From his position on a hill outside town, he could see that early seizure of the high ground west of Gettysburg was critical to giving the Army of the Potomac time to mass its forces. Occupation of this high ground would also preserve the tactical advantage of the high ground to Buford's rear for the Union Army once they arrived on the battlefield. Buford also knew that if the Confederates were allowed to mass their forces

**Visualizing the Battle:** Gettysburg, 1863

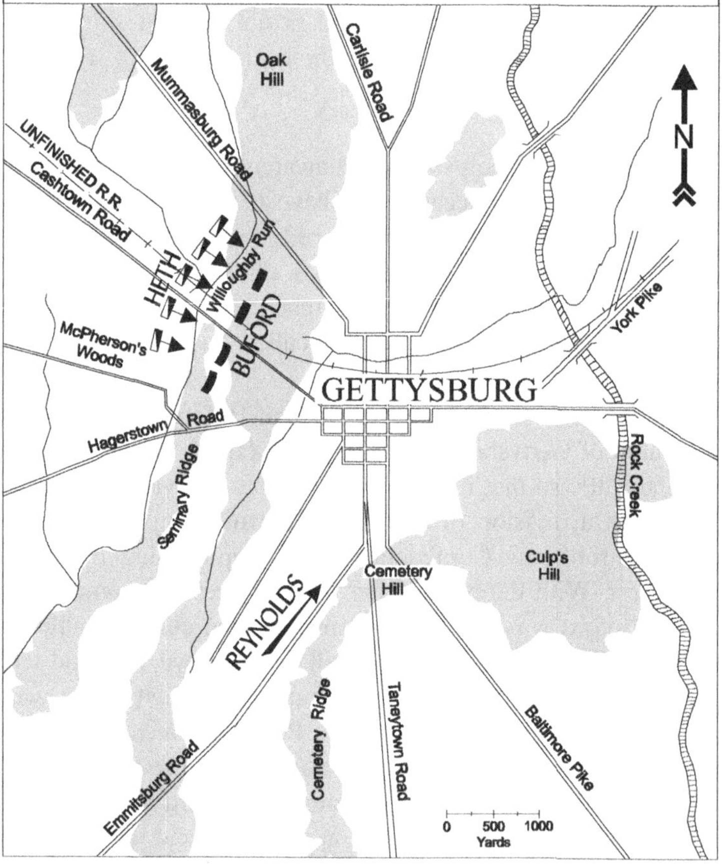

first around the high ground to the south and west, Lee would have the advantage over the arriving Union forces. (See figure.)

Quickly spreading out one brigade west of town along McPherson Ridge, General Buford settled in to defend Gettysburg until the arrival of Union reinforcements. On July 1st, the following day, he held his ground against a division of Confederate infantry supported by artillery until General John Reynolds' Second Corps came up and reinforced the line. General Buford's ability to foresee the coming battle, take quick action in the disposition of his forces, and hold the high ground until reinforced was one of the decisive actions that defeated the Army of Northern Virginia at the battle of Get- tysburg.[4] Buford's actions at Gettysburg demonstrated an exceptional ability to grasp the essence of a tactical situation through the skills of pattern recognition and intuitive deci- sionmaking.

**Acting Decisively**

Our ability to understand the situation is useless if we are not prepared to act decisively. When the opportunity arrives, we must exploit it fully and aggressively, committing every ounce of combat power we can muster and pushing ourselves to the limits of exhaustion. The keys to this effort are identifying enemy critical vulnerabilities, shaping the operating area to our advantage, designating a main effort to focus our combat power, and acting in a bold and ruthless manner.

***Critical Vulnerabilities.*** For battlefield success, it is not enough to generate superior combat power. We must focus that combat power. We must concentrate our efforts on a *critical vulnerability*, that is, a vulnerability which permits us to destroy some capability without which the enemy cannot function effectively.

Seeking the enemy's vulnerabilities means striking with our strength against his weakness (rather than his strength) and at a time when the enemy is not prepared. This is where we can often cause the greatest damage at the lowest cost to ourselves. In practical terms, this often means avoiding his front, where his attention is focused, and striking his flanks and rear, where he does not expect us.

Just because a target is vulnerable does not, however, mean that it is worth attacking. We must direct our resources and strike at those capabilities that are *critical* to the enemy's ability to function—to defend, attack, or sustain himself, or to command his forces. We must focus our efforts on those critical vulnerabilities that will bend the enemy to our will most quickly.

At the lower tactical level, this may mean using fire and maneuver to take out a machine gun position that is the backbone of an enemy defense. It may mean using a gap in the enemy's fields of fire that allows us to get into the rear of his position. It may mean exploiting the enemy's lack of air defenses by calling in close air support. It may mean taking advantage of an enemy's lack of mobility by rapidly overrun- ning a key position faster than he can respond. It may mean interdicting enemy resupply routes when his supplies are running short. It may mean exploiting a lack of long-range weapons by employing standoff tactics. Whatever we determine the enemy's critical vulnerability to be, we must be prepared to rapidly take advantage of it.

There is no formula for determining critical vulnerabilities. Each situation is different. Critical vulnerabilities will rarely be obvious. This is one of the things that make mastery of tactics so difficult and one reason that so few actions achieve a decisive outcome. Identifying critical vulnerabilities is an important prerequisite to achieving a decision.

***Shaping the Operating Area.*** Once we have developed an understanding of the situation and have determined enemy critical vulnerabilities to attack, we try to shape the operating area to our advantage. Shaping includes both lethal and nonlethal activities such as planning fires to fix the enemy, using an axis of advance to facilitate movement, designating objectives to focus our combat power, or using deceptive measures to reinforce enemy expectations. Shaping activities can make the enemy vulnerable to attack, impede or divert his attempts to maneuver, facilitate the maneuver of friendly forces, and otherwise dictate the time and place for decisive battle. Shaping forces the enemy to adopt courses of action favorable to us. We attempt to shape events in a way that allows us several options, so that by the time the moment for decisive action arrives, we have not restricted ourselves to only one course of action. Through shaping we gain the initiative, preserve momentum, and control the tempo of combat.

***Main Effort.*** The main effort is a central maneuver warfare concept: concentrating efforts on achieving objectives that lead to victory. Of all the actions going on within our command, we recognize one as the most critical to success at that moment. The unit assigned responsibility for accomplishing this key mission is designated as the main effort—the focal point upon which converges the combat power of the force.

The main effort receives priority for support of any kind. It must be clear to all other units in the command that they must support that unit in the accomplishment of its mission. The main effort becomes a harmonizing force for a subordinate's initiative. Faced with a decision, we ask ourselves: *How can I best support the main effort?*

Some actions may support the main effort indirectly. For example, a commander may use other forces to deceive the enemy as to the location of the main effort. Marine forces used this concept extensively in conducting a series of combined arms raids prior to the ground offensive in Operation Desert Storm. The raids were to confuse the Iraqis as to the true position and intention of Allied forces. "The raid force appeared in the middle of the night and fired from positions the enemy had every right to believe were unoccupied."[5]

Use of a main effort implies the use of *economy of force.* This term does *not* mean that we use as little force as we think we can get away with. Rather, it means that we must not fail to make effective use of all of the assets available to us. Forces not in a position to directly support the main effort should be used to indirectly support it. Such forces might be used to distract the enemy or to tie down enemy forces that might otherwise reinforce the threatened point. Uncommitted forces can be used in this effort by maneuvering them in feints and demonstrations that keep the enemy off balance.

While a commander always designates a main effort, it may shift during the course of a battle as events unfold. Because events and the enemy are unpredictable, few battles flow exactly as the commander has planned. As a result, the commander must make adjustments. One way is by redesignating the main effort. For example, if Company A is desig- nated as the main effort but runs into heavy enemy resistance while the adjacent Company B makes a breakthrough that exploits a critical vulnerability, the battalion commander may designate Company B as the main effort. This new designation of Company B as the main effort must not, however, be merely nominal. It means that the combat power which was supporting Company A now shifts to support Company B.

Identifying the main effort is the principal and most important answer to the question, "How do we achieve a decision?"

***Boldness and Ruthlessness.*** Forcing a successful decision requires the commander to be bold and ruthless. Boldness refers to daring and aggressiveness in behavior. It is one of the basic requirements for achieving clear-cut outcomes: In order to try for victory, we must *dare* to try for victory. We must have a desire to "win big," even if we realize that in many situations the conditions for victory may not yet be present. Ruthlessness refers to pursuing the established goal mercilessly and single-mindedly. This is doubly important once we gain an advantage. Once we have an advantage, we should exploit it to the fullest. We should not ease up, but instead increase the pressure. Victory in combat is rarely the product of the initial plan, but rather of ruthlessly exploiting any advantage, no matter how small, until it succeeds.

Boldness and ruthlessness must be accompanied by strong leadership and tempered by sound judgment. Without these qualities, boldness can become recklessness, and ruthlessness can be distorted into cruelty.

## CONCLUSION

As Marine leaders, whether of fire teams or of a Marine expeditionary force, we are responsible for achieving success. In combat, the success we seek is victory—not merely a partial or marginal outcome that forestalls the final reckoning, but a victory that settles the issue in our favor.

To be victorious, we must work ceaselessly in peacetime to develop in ourselves a talent for military judgment—the ability to understand a situation and act decisively. Military judgment results from the wisdom gained from experience. It allows us to identify patterns of activity and to concentrate our efforts against a critical vulnerability that will bend the enemy to our will. We must sharpen our ability to make decisions intuitively based on our understanding of the situation.

# Chapter 3

# Gaining Advantage

*"In war the power to use two fists is an inestimable asset. To feint with one fist and strike with the other yields an advantage, but a still greater advantage lies in being able to interchange them—to convert the feint into the real blow if the opponent uncovers himself."*[1]

—B. H. Liddell Hart

*"The challenge is to identify and adopt a concept of warfighting consistent with our understanding of the nature and theory of war and the realities of the modern battlefield. What exactly does this require? It requires a concept of warfighting that will function effectively in an uncertain, chaotic, and fluid environment—in fact, one that will exploit these conditions to advantage."*[2]

—FMFM 1, *Warfighting*

A basic principle of martial arts is to use the opponent's strength and momentum against him to gain more leverage than one's own muscles alone can generate, thereby gaining an advantage. The same concept applies to tactics. We strive to gain an advantage over our adversary by exploiting every aspect of a situation to help us to achieve victory, not by overpowering him with our own strength. This chapter will discuss several different ways of generating leverage to gain advantage over the enemy.

Consider the American Indian ambush technique. A small number of warriors would draw a superior force of pursuing cavalry into a canyon or similar close terrain. There a larger force of warriors, lying in wait, would quickly surround and ambush the soldiers, who thought they had been pursuing a retreating enemy. By exploiting the cavalry's initial advantages of strength and momentum, the American Indians were able to seize the initiative and gain the advantage through the use of this classic ambush method.

## Combined Arms

The use of combined arms is a key means of gaining advantage. It is based on the idea of presenting the enemy not merely with a problem, but with a *dilemma*—a no-win sit-uation. We combine supporting arms, organic fires, and maneuver in such a way that any action the enemy takes to avoid one threat makes him more vulnerable to another.[3] For example, an entrenched enemy should discover that if he stays hunkered down in fighting holes, Marine artillery and air will blast him out. If he comes out to attack, Marine infantry will cut him down. If he tries to retreat, Marine armor and airpower will pursue him to his destruction. *That* is combined arms.

A good example of the use of combined arms at the squad level would be the squad leader positioning squad automatic weapons and grenade launchers to provide support by fire while infantrymen with rifles assault the position. The firepower from the automatic weapons keeps the enemy in their fighting holes while grenades make those holes untenable. These supporting fires keep the enemy from reacting effectively to our maneuvering infantry force. The enemy forces are placed in a no-win situation.

Modern tactics is combined arms tactics. That is, it combines the effects of various arms—infantry, armor, artillery, and aviation—to achieve the greatest possible effect against the enemy. Artillery and infantry, for example, are normally employed together because of their mutually reinforcing capabilities—the infantry provides close support to the artillery, protecting them from dismounted threats, while the artillery provides the infantry with timely, close, accurate, and continuous fire support. The strengths of the arms complement and reinforce each other. At the same time, the weaknesses and vulnerabilities of each arm are protected or offset by the capabilities of the other.

While a division commander in 1941, General Patton had the following comments regarding combined arms:

> There is still a tendency in each separate unit . . . to be a one-handed puncher. By that I mean that the rifleman wants to shoot, the tanker to charge, the artilleryman to fire . . . . That is not the way to win battles. If the band played a piece first with the piccolo, then with the brass horn, then with the clarinet, and then with the trumpet, there would be a hell of a lot of noise but no music. To get harmony in music each instrument must support the others. To get harmony in battle, each weap- on must support the other. Team play wins.[4]

The Marine air-ground task force is a perfect example of a balanced combined arms team. Combined arms tactics is standard practice and second nature for all Marines.

## Maneuver

Maneuver provides us a means to gain an advantage over the enemy. In too many battles, one or both sides have sought to gain advantage in combat through firepower and attrition. In World War I, one side would rush across no-man's-land under murderous fire and attempt to push an opponent off desired terrain. If the attack succeeded—and few did—the evicted forces counterattacked in the same manner, usually reoccupying the same terrain they had before. These battles were firepower and attrition contests, and the advantage lay with the side that had the most personnel and equipment to expend. The cost in casualties and equipment was high and often produced no decisive results. We want to avoid this type of engagement.

Traditionally, maneuver has meant moving in a way that gains positional advantage. For example, we may maneuver by enveloping an exposed enemy flank or by denying the enemy terrain critical to his goals. We may maneuver by threatening the enemy's lines of communications and forcing him to withdraw. We may maneuver by seizing a position which allows us to bring effective fire to bear against the enemy but which protects us against enemy fires. We may maneuver in other dimensions as well. For instance, we may also maneuver in time by increasing relative speed and operating at a faster tempo than the enemy. Normally we maneuver both in time and space to gain advantage and, ultimately, victory at the least possible cost.

## Exploiting the Environment

The use of the environment offers tremendous opportunities to gain advantage over the enemy. We must understand the characteristics of any environment where we may have to operate:

jungle, desert, mountain, arctic, riverine, or urban. More importantly, we must understand how the effects of terrain, weather, and periods of darkness or reduced visibility impact on our own and our adversary's ability to fight.

**Terrain**

Our objective is to employ tactics that makes terrain an advantage to us and a disadvantage to our opponent. Terrain impacts on our maneuver and influences our tactical disposi- tions. We must understand terrain and comprehend its effects, as it may limit our movement, reduce our visibility, or restrict our fires. We must understand what effects it has on the enemy and on his abilities to detect or engage us. We must be aware that the enemy also seeks advantage from terrain. We must understand that terrain shapes the enemy's maneuver and dispositions as well as our own.

Lieutenant Harrol Kiser of the 1st Battalion, 7th Marine Regiment, knew how to use terrain to gain an advantage. In November 1950, his company was ordered to seize a key piece of terrain at Toktong Pass during the march out of the Chosin Reservoir area. Lieutenant Kiser had only 20 Marines left in his platoon, and the pass was heavily defended by the Chinese. Using a flanking ridgeline to conceal his approach, Lieutenant Kiser skillfully enveloped the enemy from the rear and quickly routed the Chinese out of their well-entrenched position.[5] Today, as in Korea, the intelligent use of terrain has become a standard practice for Marines.

**Weather**

Adverse weather—cold, heat, rain—impedes combat operations. The military unit that is best prepared to operate in these conditions will gain an advantage over its opponent. During the breakout from Chosin Reservoir in November 1950, Marines demonstrated time and time again the ability to use harsh weather to their advantage over a determined enemy. The assault of Able Company, 1st Battalion, 1st Marine Regiment, on Hill 1081 in a blinding snowstorm is such an example. Despite visibility of only 25 yards, the company was able to coordinate a combined arms attack and envelop this key piece of terrain that blocked the breakout of the 1st Marine Regiment. Using a snowstorm to mask its movement, Able Company surprised and annihilated the Chinese defenders, thereby opening a route for the rest of the division.[6]

If we are to use weather to our advantage, we must train and prepare rigorously to operate in all climatic conditions. We must be able to operate our equipment and employ our weapons effectively in hot, cold, or wet environments—literally in every clime and place.

**Periods of Darkness or Reduced Visibility**

Units that can operate effectively during hours of darkness or periods of reduced visibility often gain significant advantage over their opponent. Reduced visibility can make the simplest of tasks difficult to accomplish. This obvious disadvantage can be turned on its head and used to our advantage by a commander whose forces are trained, equipped, able, and willing to operate at night. Night operations can produce great gains against a force that cannot or will not operate at night. Operating during periods of reduced visibility creates tempo by adding another 10 to 12 hours to the day for fighting. The psychological impact of night fighting is also great and can produce significant rewards.

A good example of the tactical impact of night attacks is found in the battle for Okinawa during World War II. Marine forces were essentially stalemated by the presence of a strong Japanese defensive line in the coral ridges of southern Okinawa. After days of ineffective attacks by the 7th Marine Regiment, the regimental commander elected to attack under cover of darkness. At 0330 on 12 June 1945, the 1st and 2d Battalions of the 7th Marines advanced, using a road that intersected the ridge as a guide. Colonel Edward W. Snedecker, Commanding Officer of the 7th Marines at the time, noted:

> . . . two companies, one from each [of] the 1st and 2d Battalions, got across the valley during the night into position [on the ridge]. Early in the morning when the Japanese came out to cook breakfast, they found a little bit of a surprise . . . [for] them.[7]

The Japanese defenders were not used to U.S. forces attacking at night. The use of darkness allowed Marines to occupy positions along the crest of Kunishi Ridge literally without firing a shot. From these positions, the Marines dislodged the enemy from their entrenched positions and moved onward until the Japanese defenders were annihilated.[8]

## Complementary Forces

Complementary forces—the idea of fix-and-flank—are an important way of gaining advantage. The idea behind complementary forces is to use our forces as a nutcracker. We seek to crush the enemy between two or more actions. Consider the case of an enemy rifleman firing from behind a tree. If one Marine fires from the front, the enemy rifleman is protected by the tree. If the Marine maneuvers and attempts to fire from behind, the enemy rifleman merely moves to the other side of the tree to maintain his protection. However, two Marines can place our opponent in a dilemma. One can fire from the front while the other sneaks around and fires at the enemy from the flank or rear. The opponent is now vulnerable to one or the other of the two Marines. He cannot use the tree for protection against both.

The same idea applies in air-to-air tactics. Upon detecting enemy aircraft, a flight of fighters splits into two or more elements beyond air-to-air missile range. They approach the enemy aircraft from multiple directions and varying altitudes. No matter how the enemy aircraft moves—dives, climbs, turns, or twists—it is exposed.

Sun Tzu described this concept as the *cheng* and the *ch'i*.[9] The *cheng* is the more direct, obvious action. It fixes the enemy. The *ch'i* is the unexpected or extraordinary action. It is the bid for a decision, or, as we call it today, the main effort. These two actions work together against the enemy. The two actions are inseparable and can be interchangeable in battle; the *cheng* may become the *ch'i*. The concept is basic, but it can be implemented in a variety of combinations limited only by our imagination.

## Surprise

Achieving surprise can greatly increase leverage. In fact, surprise can often prove decisive. We try to achieve surprise through deception, stealth, and ambiguity.

"War is based on deception,"[10] stated Sun Tzu. We use deception to mislead our opponents with regard to our real intentions and capabilities. By employing deception, we try to cause our opponents to act in ways that will eventually prove prejudicial for them. We may use deception to mislead the enemy as to the time and location of our pending attack. We may use deception to create the impression that our forces are larger than they really are. We hope the enemy will realize this deception only when it is too late for them to react.

Marines have often relied on deception to mislead the enemy in regard to the location of amphibious landings. Marines used deception to create the illusion of force where there was none in Operation Desert Storm. Lieutenant General Boomer stated the situation which necessitated an extensive deception operation: "We're taking on 11 Iraqi divisions with two Marine divisions. Our force ratios are horrible. We don't want him to know that. . . ."[11] The Marines created Task Force Troy: 460 Marines imitated the activities of a 16,000-man division using loudspeakers, dummy tanks and artillery, and helicopters conducting simulated resupply.

Surprise can be generated through stealth. Stealth is used to advantage when maneuvering against an enemy. It provides less chance of detection by the enemy, leaving him vulnerable to surprise action for which he may be unprepared. Marines may also employ stealth by lying in wait for an approaching enemy—an ambush. The ambush is perhaps the most effective means of surprising opponents, especially at the lower tactical level where surprise through stealth is easiest to achieve.

We can also achieve surprise through ambiguity. It is usually difficult to conceal all our movements from the enemy, but we can sometimes confuse him as to the meaning of what he sees. Sun Tzu said:

> The enemy must not know where I intend to give battle. For if he does not know where I intend to give battle he must prepare in a great many places. And when he prepares in a great

many places, those I have to fight in any one place will be few.[12]

Ambiguity was central to the tactics of the World War II German *blitzkrieg*. An attack in *blitzkrieg* involved multiple thrusts with reinforcements following whichever thrusts were most successful. The multitude of thrusts created paralyzing uncertainty because the opponent could not determine which constituted the real attack. There was nothing secret about the German attack, but it was ambiguous on a massive scale.

## Trapping the Enemy

Modern tactics is based not on pushing the enemy, but on trapping him—another excellent way of gaining advantage. Trapping is the desired result of the application of combined arms, fire and maneuver, or complementary forces tactics.

Why do we want to trap the enemy instead of just push him? A pushing contest is seldom decisive. The side that is pushed out comes back the next day still full of fight. We have to fight him again and again. Unfortunately, in Vietnam, many of our battles were pushing battles. We were always able to push the enemy off the ground he held and to inflict casualties on him. He just withdrew, regrouped, replaced his losses, and came back to fight us again. The result was a series of indecisive actions and a seemingly endless war.

However, if we can trap our enemy, we have a better opportunity to win decisively. Many of history's decisive battles have been trapping actions. Recall how the Roman legions were trapped at Cannae or the German divisions at Stalingrad? Trapping gains advantage by disrupting the enemy's mental process while he attempts to think through the dilemma we have placed him in. Trapping allows us to gain and maintain the initiative as the enemy is forced to react to our actions. It can also temporarily undermine the enemy's will to resist when he is at his weakest—while we continue to press the attack and our initiative.

A good example of trapping from the Vietnam conflict occurred during Operation Dewey Canyon. (See figure.) North Vietnamese activity along the Laotian-South Vietnamese border increased dramatically in early January 1969. Large enemy convoys, including armored vehicles, regularly traveled from Laos into South Vietnam, threatening friendly units. Colonel Robert H. Barrow and his 9th Marines responded with Operation Dewey Canyon.

The three battalions of the 9th Regiment crossed the Da Krong River on February 11th and 12th. The Third and First Battalions moved south-southeast through the mountainous terrain toward Laos. Second Battalion, to the west, swung south-southwest, turning east astride the south Vietnam-Laos

**Trapping the Enemy:** Vietnam, 1969

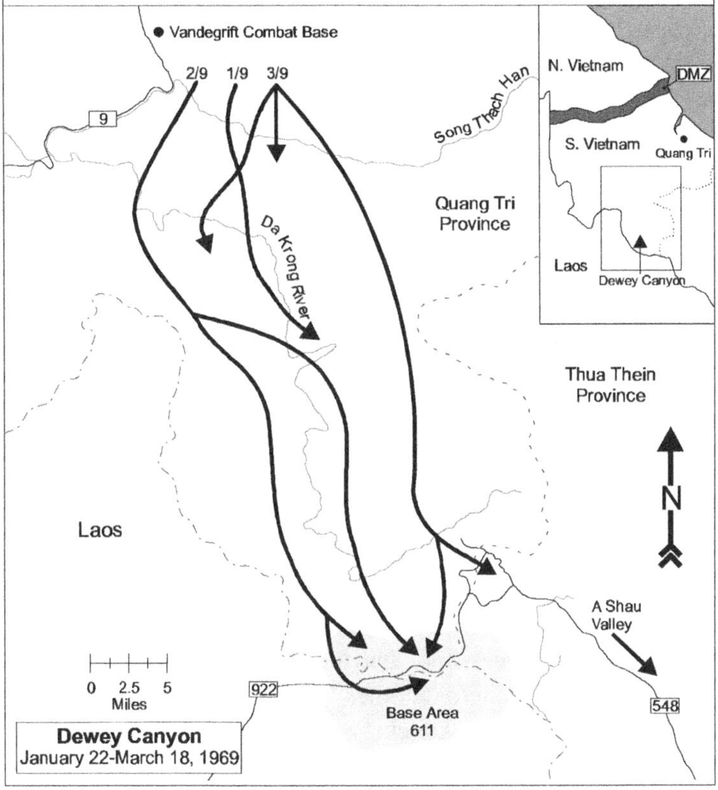


border. The North Vietnamese forces moving along Route 922 from Laos into the A Shau Valley were trapped between the three battalions. The North Vietnamese were mauled as a result. Their equipment losses were staggering. More importantly, Operation Dewey Canyon destroyed a North Vietnamese base area and so disrupted their logistics that it forced them to abandon their planned spring offensive in I Corps' area.[13]

## Developing an Ambush Mentality

Perhaps the most common tactical tool for gaining advantage is the ambush. All Marines are familiar with an ambush as a type of combat patrol.[14] In maneuver warfare, ambush takes on a much broader meaning, and the *development of the ambush mentality* is integral to maneuver warfare tactics.

The ambush mentality is probably not new to most of us. We may have employed the ambush mentality in sports. In football, the trap block is an ambush. A player pulls an offensive lineman off the line, leaving a hole. When a defender comes through the hole, another lineman suddenly blocks him from the side, usually knocking him down. The players have blind-sided him. That is the ambush mentality.

In basketball, setting up a pick is an ambush. As one teammate drives to the basket, another steps into the defender's path from behind, blocks the path, stops the defense, and momentarily clears the lane to the basket for the other teammate. Again, that is the ambush mentality.

In combat, we move our reinforced squad into position along a well-traveled trail. We position flank security to protect ourselves and give identification and warning of enemy movements down the trail. We position our weapons so as to concentrate our fires into a "kill zone" and to seal off exits, forcing the enemy to remain subject to our fires. The squad waits in position until signaled when they immediately respond with concentrated, sustained fires on enemy forces trapped in the kill zone. The enemy, surprised into inaction, unsure of what to do or where to move, is annihilated. Fires are maintained until all the enemy are killed or until signaled to stop. That is the ambush mentality.

The ambush mentality tries to turn every situation into an ambush. In this broader context, an ambush has several distinct features.

First, in an ambush we try to *surprise the enemy*. Think of a patrol that we ambush. Our enemies are walking through the woods when suddenly, out of nowhere, they are under fire from multiple directions. They are taking heavy casualties. The psychological impact of surprise may paralyze their thoughts and actions, leaving them incapable of reacting effectively. To have an ambush mentality means we always try to surprise the enemy, to do the unexpected. Surprise is the rule rather than the exception.

Second, we want to *draw our enemy unknowingly into a trap*. This will often involve deceiving him. We make one course of action appear inviting. When he takes that course of action, we are waiting for him.

Third, an ambush is *invisible*. If the ambush is not invisible, it ceases to be an ambush and instead becomes a target for the enemy. Whether we are defending or attacking, the enemy must not detect us until it is too late for him to react. Surprise often depends upon invisibility. That invisibility may be provided through stealth in movement or in focusing the enemy's attention elsewhere to allow our forces to maneuver without detection.

The reverse slope defense is an example of using invisibility to spring an ambush. The enemy does not know we are there until he comes over the crest of a hill and is hit by our fires. His vehicles are hit on their soft underbellies. His troops are fully exposed to our weapons. Because he could not see us until the last moment, he could not call in artillery fire on our position. The reverse slope not only protects us from his direct fire; it protects us from his observation and thus his indirect fire. That is part of the ambush mentality: *Do not let yourself be seen.*

Fourth, in an ambush we want to *shock the enemy*. Instead of taking him under fire gradually with a few weapons at long range, we wait until he is within easy range of every weapon. We then open up suddenly, all at once, with everything we have. He is paralyzed by the shock. He cannot react. Everything was going fine, and suddenly he is in a firestorm with people falling all around him. Often he will panic, making his problem worse as he reacts rather than acts.

Combined arms may be used to ambush the enemy. Artillery raids that reach deeper into his vital areas than expected can produce that same desired shock effect as a ground-based ambush. We place him in a dilemma as he attempts to move from the effects of artillery and goes right into an attack by air.

Finally, in the ambush mentality, we *always focus on the enemy*. The purpose of an ambush is not to hold a piece of terrain. It is to destroy the enemy. We use terrain to effect the ambush, but terrain itself is not what we are fighting for.

## ASYMMETRY

Fighting asymmetrically means gaining advantage through imbalance, applying strength against an enemy weakness. Fighting asymmetrically means using dissimilar techniques and capabilities to maximize our own strengths while exploiting enemy weaknesses. Fighting asymmetrically means fighting the enemy on our terms rather than on his. By fighting asymmetrically, we do not have to be numerically superior to defeat the enemy. We only have to be able to exploit his vulnerabilities.

For example, using tanks to fight enemy tanks, infantry to fight enemy infantry, and air to fight enemy air is symmetrical. Using attack helicopters to fight enemy tanks and close air support against enemy infantry are examples of fighting asymmetrically. In these examples, we gain the advantage of the greater speed and mobility of the aircraft relative to the enemy. Ambushing tanks with attack helicopters in terrain which hampers tank maneuver provides even more effect and generates even more advantage.

## Conclusion

Combat is a test of wills where the object is to win. One way to win is to gain and exploit every possible advantage. This means using maneuver and surprise whenever possible. It means employing complementary forces and combined arms. It means exploiting the terrain, weather, and times of darkness to our advantage. It means trapping our enemy by ambush or by some other means. It means fighting asym- metrically to gain added advantage. This is what Sun Tzu meant when he wrote: "Therefore a skilled commander seeks victory from the situation and does not demand it of his subordinates."[15]

# Chapter 4

# Being Faster

*"Hit quickly, hit hard and keep right on hitting. Give the enemy no rest, no opportunity to consolidate his forces and hit back at you."*[1]

—Holland M. Smith

*"For the infantryman to be truly effective . . . he will have to be as light of foot as he is quick of thought. . . . Mobility is needed most of all in the clash of arms. Swift and agile movement plus rapidity and intelligent tactical flexibility are its true essentials."*[2]

—John A. English

Usually, to think of weapons means to think of a personal rifle or pistol; the unit's machine guns and mortars; or the aircraft's missiles, bombs, or guns. A logistician may realize that weapons include trucks, bulldozers, and excavators. Some Marines overlook one of their most powerful weapons, one that creates advantage for infantrymen, aviators, and logisticians equally. That weapon is speed.

## Speed in Combat

How is speed a weapon? Think of sports again: The breakaway in hockey uses speed as a weapon. By rapidly passing the puck down the ice, one team denies the other the chance to set up a defense. Speed circumvents their opponent's ability to respond in an organized manner. The fastbreak in basketball seeks the same result. In two or three passes, the ball is downcourt and the basket scored, all before the opposition can re- act.

The results of speed often reach beyond the immediate goal. How many times have we seen a team score on a fastbreak, steal the ball as it comes inbounds, and immediately score again, and even a third time? Unable to regain their composure, the victims of the fastbreak become the victims of a rally. The victims lose confidence. Passes go astray; signals become crossed; tempers flare; arguments ensue. The rally becomes a rout. The beleaguered players see certain defeat. They virtually give up while still on the court.

The same thing can happen in combat. The battalion or fighter aircraft or logistics train that can consistently move and act faster than its enemy has a powerful advantage.

In June of 1943, during the battle of Saipan, the aggressive, hard-hitting tactics of General Holland Smith proved to be singularly successful in defeating the Japanese defenders. General Smith's tactical plan for Saipan called for applying "unremitting pressure on the enemy and . . . bypassing strong points of resistance for mopping up by reserve elements in order to press the attack to better ground."[3] Long indoctrinated with the value of speed in amphibious operations, General Smith's bypassing tactics placed the Japanese remaining in their fixed defenses at an extreme tactical disadvantage. These tactics proved very effective in isolating and reducing the Japanese defense. General Smith's use of speed served as a force multiplier, and it also reduced Marine casualties.

The British Royal Air Force bested the Germans during the Battle of Britain in World War II in part because they were able to speedily recover their downed pilots, return them to base, place them in new aircraft, and have them fighting again in the afternoon. Downed German pilots were less easily recovered, and the Luftwaffe had fewer of the long-range air- craft required for replacement. Eventually, pilot and aircraft losses forced the Germans to end daylight bombing and resort strictly to relatively ineffective night attacks.

Great leaders have repeatedly stated the value of speed in combat. Napoleon said, "I may lose a battle, but I shall never lose a minute."[4] Nathan Bedford Forrest told the secret of his many victories: "Get there first with the most men."[5] General Patton said in 1943, "When the great day of battle comes remember your training and remember above all else that speed and violence of attack are the sure road to success."[6] History's great commanders differed in many ways, but one thing they shared was a sense of the importance of speed.

In Operation Urgent Fury in 1983, the Marines of Battalion Landing Team 2/8, moved fast, as their commander, Lieutenant Colonel Ray Smith, had trained them to do. When they captured the operations officer of the Grenadian army, he said to them, "You appeared so swiftly in so many places where we didn't expect you that it was clear that resistance was hopeless, so I recommended to my superiors that we lay down our arms and go into hiding."[7] That is what speed used as a weapon can do for you.

## WHAT IS SPEED?

"What is speed?" would seem to have a simple answer: speed is going fast. This is speed as we think of it when driving a car—more miles per hour.

That is part of the answer in tactics as well. We use speed to gain the initiative and advantage over the enemy. For example, when a tank battalion attacks, it goes over the ground as fast as it can. General Balck was asked whether the Russian tanks ever used terrain in their attacks against him in World War II. He replied that they had used terrain on occasion, but that they more often used speed. The questioner followed up: "Which was harder to defend against?" Balck answered, "Speed."[8]

Physical speed, moving more miles per hour, is a powerful weapon in itself. On our approach to the enemy, speed in movement reduces his reaction time. When we are going through him or around him, it changes the situation faster than he can react. Once we are past him, it makes his reaction irrelevant. In all three cases, speed impacts on the enemy, especially his mind, causing fear, indecision, and helplessness. Remember, attacking the enemy's mind is a central tenet of maneuver warfare.

## Speed and Time

In a military sense, there is more to speed than simply going fast, and there is a vital difference between acting rapidly and acting recklessly. With time we must always consider the closely related factor of timing. Speed and time are closely related. In fact, speed is defined in terms of time: miles or kilometers per hour. In tactics, what this means is that time is always of the utmost importance. Time that cannot be spent in action must be spent thinking about how to act effectively.

Even when we are engaged with the enemy, we are not always moving fast. Some of the time we are not moving at all. Nonetheless, every moment is still of the utmost importance even when we are sitting still. A battalion staff that takes a day to plan an action is obviously slower than one that takes an hour. A tank battalion that takes 3 hours to refuel is slower than one that takes 2 hours, just as one that must refuel every hundred miles is slower than one that must refuel every two hundred. A company that sits down to eat once it has taken its objective is slower than one that immediately presses on into the enemy's depth. A fighter squadron that can fly only three sorties per aircraft per day is slower, in terms of effect on the enemy, than one that flies six. A maintenance repair team that takes 2 days to fix a damaged vehicle and get it back into action is slower, in terms of effect on the enemy, than one that can do it overnight.

Making maximum use of every hour and every minute is as important to speed in combat as simply going fast when we are moving. It is important to every member of a military force whether serving on staffs or in units—aviation, combat service support, ground combat, everyone. A good tactician has a constant sense of urgency. We feel guilty if we are idle. We never waste time, and we are never content with the pace at which events are happening. We are always saying to ourselves and to others, "Faster! Faster!" We know that if speed is a weapon, so is time.

## Timing

We employ speed and use time to create tempo. Tempo is not merely a matter of acting fastest or at the earliest opportunity. It is also a matter of timing—acting at the right time.

Timing requires an appreciation for the rhythm of combat so we can exploit that rhythm to our advantage. It is physically impossible to operate always at peak tempo. Even though we can extend operating cycles through the economical use of resources, we cannot operate at top speed indefinitely. We must rest our people and replenish our supplies. The test of skill is to be able to generate and maintain a fast pace when the situation calls for it and to recover when it will not hurt us.

Timing means knowing when to act and, equally important, when *not* to act. Although speed is an important tactical weapon, there are situations in which it is better to bide our time. If our concept of operations involves a diversion, we need to allow time for the diversion to take effect. If we have laid an ambush for the enemy, we need to give the enemy time to fall fully into the trap. If a situation is still forming, we may want to develop it further before we commit to a course of action. For example, an error commonly made by defenders is counter-attacking too soon so that the enemy is merely pushed back rather than cut off, encircled, and destroyed. Decisive action is our goal, and it must be timed to occur at the proper moment. There are times to act, and there are other times to set the stage and wait.

A benefit from a decision not to act is that it saves precious resources and energy for later commitment. Some leaders dissipate their units' energy on constant, unprioritized activity. Not all activities support the mission. A unit's energy is not easily replenished and should be treated as a precious resource to be expended only towards decisive goals.

## RELATIVE SPEED

Going fast and making efficient use of time are both parts of the answer to the question, "What is speed?" However, something else must be considered: the enemy. As with all things in war, speed is relative. Speed is meaningful militarily only if we are acting *faster than the enemy.* We can do that either by slowing the enemy or by increasing our own speed.

In the battle for the Falkland Islands in 1982, the British Army moved slowly. The terrain was difficult, the weather was abominable, and much of the material had to be moved on men's backs, all of which slowed down the British. Nevertheless, the British still had the advantage in speed because they moved faster than the Argentines who, once they had made their initial dispositions, essentially did not move. That superiority in relative speed allowed the British to maintain the initiative throughout the campaign.

## Continuing Speed

To be consistent, superiority in relative speed *must continue over time*. It is not enough to move faster than the enemy only now and then because when we are not moving faster, the advantage, the initiative, passes to him. Most forces can manage an intermittent burst of speed but must then halt for a considerable period to recover between bursts. During that halt, they are likely to lose their advantage. We realize that we cannot operate at full speed indefinitely, and the challenge is to be consistently faster than the enemy.

One way to sustain speed is to use the effects of combined arms. When the infantry or mounted troops must break contact temporarily to maneuver, resupply, or recover, air or artillery can keep the pressure on. Maneuver cannot be sus- tained indefinitely, but the momentum can be maintained through skillful planning of combined arms effects, keeping the enemy always at a disadvantage.

Here the speed of logistics becomes critical. Although physical exhaustion is a factor, halts often are driven by logistics: ground or aviation units must stop for equipment repair, maintenance, and resupply. Supporting forces can minimize loss of speed if they can deliver the supplies and perform the maintenance quickly. Thus, they enable combat units to move before the enemy gains the initiative.

## Speed and Change

In order to act consistently faster than the enemy, it is necessary to do more than move quickly. It is also necessary to make *rapid transitions* from one action to another. While there are many types of transitions in combat, the important thing to remember is that transitions produce friction. Reduction of friction minimizes the loss of tempo that the friction generates at the point of transition. A unit that can make transitions faster and more smoothly than another can be said to have greater relative speed.

In the 18th century, the importance of fast transitions (sometimes called agility) was displayed when shifting from column formation into line. If an army could not rapidly de- ploy into line and consequently was engaged while still in column, it was often beaten. Much drill was devoted to practicing this difficult transition so that it could be accomplished rapidly in combat. Today we develop proficiencies in battle drills and immediate-action drills that allow units to rapidly transition from one formation to another without pausing.

It is important to be able to effect rapid changes in organization as well. Being quick to effect required changes in task organization based on a rapidly changing battle situation increases agility and decreases reaction times. Battle drills and rehearsals can be conducted to smooth out procedures for changing organization rapidly. The faster these transitions can be made, the more effective the force becomes.

The place in time and space where transitions occur can be called a *friction point.* Friction points commonly encountered in tactics include movement from an assembly area to attack; from patrol movement formation to ambush posture; from defensive posture to attack; from one maneuver to another, and so forth. The transition involves simply positional changes and drills, but also changes of attitude in the minds of Marines. We must shift our mental focus from one movement to another.

A modern example of the importance of fast transitions comes from aerial combat. In the Korean War, American aviators achieved a high kill ratio of about 10:1 over their North Korean and Chinese opponents. At first glance, this is somewhat surprising. The main enemy fighter, the MiG-15, was superior to the American F-86 in a number of key respects. It could climb and accelerate faster, and it had a better sustained turn rate. The F-86, however, was superior to the MiG in two critical, though less obvious, respects. First, because it had high-powered hydraulic controls, the F-86 could shift from one maneuver to another faster than the MiG. Second, because of its bubble canopy, the F-86 pilot had better visibility. The F-86's better field of view provided better situational awareness and also contributed to fast transitions because it allowed its pilot to understand changing situations more quickly.

American pilots developed new tactics based on these two advantages. When they engaged the MiGs, they sought to put them through a series of maneuvers. The F-86's faster transitions between maneuvers gave it a time advantage that the pilot transformed into a position advantage. Often, when the MiG pilots realized what was happening, they panicked—and thereby made the American pilot's job all the easier.

These tactics illustrate the way fast transitions contribute to overall speed and to a time advantage. The importance of time and speed in a broader sense has been brought out in the work of John Boyd. A former colonel in the U.S. Air Force, Boyd studied a wide variety of historic battles, campaigns, and wars. He noted that where numerically inferior forces had defeated their opponents, they often did so by presenting the other side with a sudden, unexpected change or a series of changes. The superior forces fell victim because they could not adjust to the changes in a timely manner. Generally, defeat came at relatively small cost to the victor.[9]

This research led to the Boyd theory, which states that conflict may be viewed as time-competitive cycles of observation-orientation-decision-action (OODA). First, each party to a conflict enters the fray by observing himself, his surroundings, his enemy. In tactics, this equates to adoption of a hunting instinct: searching; actively looking; hunting for the enemy; and seeing what he is doing or is about to do. It also includes anticipating the enemy's next moves—getting inside his mind.

Second, based upon those observations, the combatant orients to the situation, that is, produces a mental image of the situation and gains situational awareness. This awareness becomes the foundation on which to erect a plan. Generally, the better the orientation, the better the plan.

Next, based upon this orientation, the combatant decides upon a course of action. The decision is developed into a plan that can be disseminated among subordinates for their planning and execution.

Last, the combatant acts, or puts the decision into effect. In tactics this is the execution phase where the decision, or plan, is implemented. Since this action has changed the situation, the combatant again observes, beginning the cycle anew. Boyd's cycle is also known as the OODA loop.

The Boyd theory helps to define the word "maneuver." It means being consistently faster than our opponent. As our enemy observes and orients on our initial action, we must be observing, orienting, deciding, and acting upon our second action. As we enact our third, fourth, and fifth move, the time gap between our actions and our enemy's reactions increasingly widens. Our enemy falls behind in a panicked game of catch up. As he tries to respond to our penetration, we attack his reserves and his command and control. As he counterattacks with his mobile reserve, we bypass with helicopterborne forces. Everything he does is too late.

Thus, the military answer to the question "What is speed?" is not simple. Nonetheless, it is central to every aspect of tactics. As General George Patton said, "In small operations, as in large, speed is the essential element of success."[10]

We should also exercise caution so as not to confuse speed with haste. General Patton made this observation:

> Haste and Speed: There is a great difference between these two words. Haste exists when troops are committed without proper reconnaissance, without the arrangement for proper supporting fire, and before every available man has been brought up. The result of such an attack will be to get the troops into action early, but to complete the action very slowly.

> Speed is acquired by making the necessary reconnaissance, providing the proper artillery [support], . . . bringing up every [available] man, and then launching the attack with a predetermined plan so that the time under fire will be reduced to the minimum.[11]

## Becoming Faster

Now we see clearly the importance of speed. We want to be fast. How do we do it?

We start by recognizing the importance of time. As leaders of Marines, we have a responsibility to make things happen fast. Our sense of the importance of time, of urgency, must direct our actions. We must work to create and build that sense within ourselves.

Once we have it, there are a number of things we can do to increase speed. First, we can *keep everything simple*. Simplicity promotes speed; complexity slows things down. Simplicity should be central to our plans, our staffs (large staffs may be one of war's greatest consumers of time), our command and control, and our own actions.

Second, speed is increased through *decentralization*. Decentralization is an important concept in the execution of maneuver warfare. How do we achieve decentralization, while still retaining control? We use two main tools that provide the required control of the effort and the decentralization of its execution. These tools are mission tactics and commander's intent.

*Mission tactics* is the assignment of a mission to a subordinate without specifying how the mission must be accomplished. It is a key tenet of maneuver warfare. In mission tactics, the higher commander describes the mission and explains its purpose. The subordinate commander determines the tactics needed to accomplish the task based on the mission and the higher commander's intent. In this way, each leader can act quickly as the situation changes without passing information up the chain of command and waiting for orders to come back down. Speed is greatly increased by this decentralization process. According to John A. English in his work *On Infantry*, decentralization has been one of the most significant features of modern war. English wrote: "In the confused and often chaotic battlefield environment of today, only the smallest groups are likely to keep together, particularly during critical moments."[12] In such circumstances, individuals rally around their leader who, armed with knowledge of the purpose or intent behind their task, can lead them toward success.

The *commander's intent* provides the overall purpose for accomplishing the task assigned through mission tactics. Although the situation may change, subordinates who clearly understand the purpose and act to accomplish that purpose can adapt to changing circumstances on their own without risking diffusion of effort or loss of tempo. Subordinate commanders will be able to carry on this mission on their own initiative and through lateral coordination with other subunits, rather than running every decision through the higher commander for approval.

A third way to become faster is through *experience*. Experience breeds speed. Experience gives units advantages over other less experienced units. This is why veteran units are usually much faster than green, untried units. If we are familiar with a situation or at least know generally what to expect, we can think, act, and move faster. In peacetime, our Marines are not likely to be combat veterans. Still, we can give them experience through tactical decision games, sand table exercises, war games, field exercises, and rehearsals. These and other forms of training help to reduce the stress and confusion of combat.

Another way in which experience helps us become faster is through the use of *implicit communications*. Implicit communications are mutual understandings that require little or no actual talking or writing. For example, two company commanders know each other well. They think alike because their battalion commander has established standing operating procedures and has schooled subordinate commanders in an approach to war. Thus, the commander of Company B does not need to talk with the commander of Company C very often in action because each knows from common past experiences and from daily observations how the other is likely to react in many different situations. If B Company's commander creates an opportunity, C Company's commander will take advantage of it. *That* is implicit communication. It is faster and more reliable than explicit communication (trying to pass words or messages back and forth over radios or telephones).

Of course, implicit communications must be developed over time. This requires actions that strengthen unit cohesion and mutual trust. This requires keeping people together in their units and stable in their assignments. It implies keeping good teams together. It means developing a band of brothers in our units, as Admiral Horatio Nelson did. He spent many evenings with his captains gathered in the cabin of his flagship talking over tactics, ways they might fight different engagements, how they would defeat this or that opponent. From those evenings came a shared way of thinking so strong that, at Trafalgar, Nelson needed only to signal "England expects every man will do his duty," and "Close action."[13] Sometimes words have meaning beyond the normally obvious meaning because of shared experiences and understanding.

Another way speed gains from experience is in the development of *lateral communication,* or coordination. If all communication is up and down the chain of command, action will move slowly. If commanders and leaders at every level communicate laterally—if we, as leaders, talk directly to other leaders—action moves much faster. Lateral communication is not a natural consequence of mission orders. It must be practiced in training. It results from the confidence of the higher commander who through past experiences has found that subordinates can exercise initiative based on the assigned mission and the commander's stated intent.

A good example of lateral communication comes from aviation. In the air, the pilots of a flight of aircraft communicate laterally as a matter of course. A pilot who needs to talk to another does so. A message need not go through the mission commander and then be relayed to the other pilot. Events would quickly outpace communication if pilots tried to talk that way. The same procedures may be employed by ground combat and logistics units as well.

A fourth way to become faster is by the commander's *positioning* himself at the point of friction. This position may be with the main effort, with a supporting effort, or in the rear. A commander who is forward can instantly influence the battle as the situation develops. For the same reason, a commander may choose a position at a crucial crossroad during a night movement, or where a unit is pushing supplies forward, or where a counterattack force in the defense may be sited. The key is to be where we can best influence the actions of our units. As Marines, we believe in leading from the front since that is where most friction points occur, but they may occur elsewhere. We must choose our positions accordingly.

Throughout World War II and his entire career, Lieutenant General Lewis B. "Chesty" Puller believed that Marines had to lead from where the fighting was. "This Command Post business will ruin the American Army and Marines if it isn't watched,"[14] he said while he was the commanding officer of 1st Battalion, 7th Marines, at Guadalcanal. As a battalion commander, Puller usually positioned himself directly behind the point element of his battalion and his headquarters element directly behind the lead company so that he could best influence the actions of his unit. From this location, he was able to impose his will and personally affect the outcome of the engagement. Depending on the situation, he could also be found at other points on the march or on his perimeter. His idea was to be where he could best influence the action.

Finally, it is important not only to be faster, but to maintain that speed through time. This endurance is made possible through *physical and mental fitness*. Physical fitness develops not only the speed, energy, and agility to move faster, but it also develops the endurance to maintain that speed for longer durations. With endurance, we not only outpace the enemy but maintain a higher tempo longer than he can. Mental fitness builds the ability to concentrate for longer periods of time and to penetrate below the surface of a problem. For this reason, fitness plays an important part in the life of every Marine. Patton once said "High physical condition is vital to victory."[15]

## Conclusion

We must be faster than our opponent. This means we must move fast, but, more importantly, we must act faster than our enemy. The aim is to tailor our tactics so that we can act faster than the enemy force can react. Our ability to plan, decide, and execute faster than our enemy creates advantage that we can exploit. We have just discussed ways to improve our speed. Readers of this publication may think of additional ways to be fast. When you find one that works, tell your fellow Marines about it so they can use it too. Anything that works to make you faster is good even if it is not yet in the books.

# Chapter 5

# Adapting

*"Victory smiles upon those who anticipate the changes in the character of war, not upon those who wait to adapt themselves after they occur."*[1]

—Giulio Douhet

*"In any problem where an opposing force exists, and cannot be regulated, one must foresee and provide for alternative courses. Adaptability is the law which governs survival in war as in life—war being but a concentrated form of the human struggle against environment."*[2]

—B. H. Liddell Hart

The modern battlefield is characterized by friction, uncertainty, disorder, and rapid change. Each situation is a unique combination of shifting factors that cannot be controlled with precision or certainty. This chapter discusses ways to think about adapting or modifying our decisions based on changed circumstances or sudden opportunities. A tactically proficient leader must be able to adapt actions to each situation.

The OODA loop discussed in chapter 4 essentially describes the process of adaptation—we observe the situation, orient to it, decide what to do, and act. The antagonist who can consistently adapt more quickly to the situation will have a significant advantage. Adaptability is thus an important part of Marine Corps tactics. In essence, adaptability means shortening the time it takes to adjust to each new situation.

There are two basic ways to adapt. Sometimes we have enough situational awareness to understand a situation in advance and take preparatory action. This is *anticipation.* At other times we have to adapt to the situation on the spur of the moment without time for preparation. This is *improvisation.* To be fully adaptable, we must be able to do both.

## ANTICIPATION

The first basic way to adapt is to anticipate, by which we mean to introduce new methods, schemes, or techniques *for future use*. In order to anticipate, we must be able to forecast future actions, at least to some extent. Our forecasts are usually based on past experiences. Often a forecast involves considering what we learned through trial and error in training, exercises, or actual combat. An excellent example of anticipation is the Marine Corps' development of amphibious warfare techniques at Quantico during the 1920s and 1930s. These techniques proved to be essential to success in World War II, both in the Pacific and in Europe.

All planning at all echelons is a form of anticipatory adaptation—adapting our actions in advance. Another important tool for tactical adaptation is the use of immediate-action drills or standing operating procedures. These are practiced, predesigned, generic actions which cover common situations. Having a collection of these tools at our disposal allows us to react immediately *in a coordinated way* to a broad variety of tactical situations. Immediate-action drills do not replace the need for tactical judgment; they merely provide a way to seize initiative in the early stages of a developing situation until we can take more considered action. They provide the basis for adaptation.

## Improvisation

The second basic way to adapt is to improvise, to adjust to a situation *on the spur of the moment without any preparation.* Like anticipation, improvisation is key to maneuver warfare. Improvisation requires creative, intelligent, and experienced leaders who have an intuitive appreciation for what will work and what will not.

Improvisation is of critical importance to increasing speed. It requires commanders who have a strong situational awareness and a firm understanding of their senior commander's intent so that they can adjust their own actions in accordance with the higher commander's desires. Often we will find ourselves in a situation where our organic resources—weapons, vehicles, and so on—are not adequate to keep us moving fast. In France in 1940, German General Heinz Guderian put some of his infantry in commandeered French buses. On Grenada, when Army Rangers needed vehicles, they took East German trucks belonging to the Grenadian army. Sound unorthodox? There is nothing "orthodox" about failure due to an inability to adapt.

For instance, take the situation in which Marines of the 2d Battalion, 5th Marine Regiment, found themselves in the battle of Hue City, Republic of Vietnam, in February 1968. One of their first objectives was to retake the city's Treasury building, which was heavily defended by the North Vietnamese. Prior to the assault, the Marines were disappointed to see that their mortar fire was having little effect on the building or its defenders. Then the battalion executive officer found some U.S. tear gas canisters and dispensers in the Military Assistance compound they had reoccupied. Realizing the North Vietnamese lacked gas masks, the Marines proceeded to lob the tear gas canisters into the Treasury building. As a result of the executive officer's quick thinking and adaptation, the North Vietnamese quickly vacated the building, and the Marines secured the objective with minimal casualties.[3]

## FLEXIBLE PLANS

We have several techniques to help us develop adaptability. One of these is to make flexible plans. Flexible plans can enhance adaptability by establishing a course of action that provides for multiple options. For example, a blocking position that covers two avenues of approach from the same location instead of only one provides the flexibility to adapt to an enemy coming through either avenue.

We can increase our flexibility by providing *branches* for current and future operations. Branches are options (e.g., changing dispositions, orientation, strength, movement, or accepting or declining battle) to deal with changing conditions on the battlefield that may affect the plan.[4]

Flexibility can also be increased by providing *sequels* for current and future operations. Sequels are courses of action to follow probable battle or engagement outcomes; victory, defeat, or stalemate.[5]

The value of branches and sequels is that they prepare us for several different actions. We should keep the number of branches and sequels to a relative few. We should not try to develop so many branches and sequels that we cannot adequately plan, train, or prepare for any of them. The skillful, well-thought-out use of branches and sequels becomes an important means of anticipating future courses of action. This anticipation helps accelerate the decision cycle and therefore increases tempo.

Flexible plans avoid unnecessary detail that not only consumes time in their development but has a tendency to restrict subordinates' latitude. Instead, flexible plans lay out what needs to be accomplished but leave the manner of accomplishment to subordinates. This allows the subordinates the flexibility to deal with a broader range of circumstances.

Flexible plans are plans that can be easily changed. Plans that require coordination are said to be "coupled." If all the parts of a plan are too tightly coupled, the plan is harder to change because changing any one part of the plan means changing all the other parts. Instead, we should try to develop modular, loosely coupled plans. Then if we change or modify any one part of the plan, it does not directly affect all the other parts.[6]

Finally, flexible plans should be simple plans. Simple plans are easier to adapt to the rapidly changing, complex, and fluid situations that we experience in combat.

## Decentralization

Another excellent way to improve adaptability is to decentralize decisionmaking authority as much as each situation allows. This means that commanders on the scene and closest to the events have the latitude to deal with the situation as required *on their own authority*—but always in accordance with the higher commander's intent. This decentralization speeds up reaction time: we do not have to wait for information to flow up to a higher commander and orders to flow back down. It increases the responsiveness of the organization, which in turn increases adaptability. Decentralizing control through the use of mission orders is one of the tools we use to maximize our ability to adapt.

Confidence in the abilities of subordinates plays an important part in decentralization. Leaders who have confidence in the capabilities of their subordinates will feel more comfortable in granting them greater latitude in accomplishing tasks. It fosters a climate where senior leaders know that their intent will be carried out. This was particularly true for the 1st Battalion, 7th Marines, during Operation Desert Storm. As the battalion began breaching operations for the advance of the 1st Marine Division across the first two Iraqi mine belts, Marines were suddenly overwhelmed with " 'hundreds upon hundreds of Iraqis sporting white flags' "[7] who were trying to surrender. The number was so great that it threatened to stop the Marine advance. However, the battalion commander immediately recognized the situation, judged that the Iraqis were harmless, and instructed the battalion not to stop to accept their surrender. "It was precisely the . . . type of local situation that [the division commander] wanted his commanders to recognize and use their own initiative to correct."[8] Here the commanding officer who was closest to the situation and who understood the division commander's intent not to lose the momentum of the advance adapted to the situation. This adaptation resulted in a rapid breach of Iraqi defenses.

## Conclusion

Successful warfare is filled with examples of leaders adapt- ing to changing situations. We must start to learn how to adapt now during our training. Leaders should value and encourage innovative thinking. Moreover, they should expect creative thinking from their subordinates because it creates new opportunities.

For adaptation to be effective, commanders must readily exploit the opportunities uncovered by subordinates. Commanders cannot remain tied to plans that blind them to fleeting opportunities. While making the best possible preparations, they must welcome and take advantage of unforeseen opportunities.

## Chapter 6

# Cooperating

*"Unity of command (effort) is coordinated action toward a common goal; it is cooperation. It is working together by all commanders toward the accomplishment of a common mission, which is imperative for complete and final success. Commanders must develop in their staffs and subordinates the desire to cooperate, not only among themselves but with other elements of the command."*[1]

—NAVMC 7386, *Tactical Principles*

*"The first element of command and control is* people—*people who gather information, make decisions, take action, communicate, and cooperate with one another in the accomplishment of a common goal."*[2]

—MCDP 6, *Command and Control*

Everything that we have to do in tactics—gaining advantage and, above all, achieving a decisive result—needs a team effort. If efforts are not in harmony, results may be indecisive. For example, if the aviation combat element's actions are not harmonized with those of the ground combat element, they are unlikely to have a decisive effect. If artillery support is not well coordinated with an infantry attack, combined arms synergy will not be achieved, and the attack may fail. However, achieving this team effort is easier said than done. It requires rapidly maneuvering forces, often widely dispersed, to work together under the most adverse conditions.

## Control in Combat

Because war is characterized by chaos, uncertainty, and rapid change, control quickly breaks down. It is probably a mistake to speak of control in combat. MCDP 6 states that "given the nature of war, it is a delusion to think that we can be in control with any sort of certitude or precision."[3] As anyone who has experienced combat will undoubtedly agree, it is impos- sible to control everything. Attempts to impose control also can easily undermine the initiative upon which Marine Corps tactics depends. Marines can become hesitant; they may feel they must wait for orders before acting. We are not likely to move faster or gain leverage over a competent opponent unless Marines at every level exercise initiative.

The dilemma, then, is this: How do we achieve the goal of working together in harmony while exercising a more decentralized type of control?

## Cooperation

The beginning of the answer lies in *cooperation.* We define cooperation as the union of self-discipline and initiative in pursuit of a common goal. Cooperation can be viewed as a component of control.

Control can generally be divided into two types: centralized and decentralized. Centralized control tends to be in one direction and works from the top down: someone at a higher level determines what subordinates will and will not do. Centralized control makes us conform to higher dictates because only one person does the thinking for the organization—the person in control.

In contrast, decentralized control works from the bottom up. Command is the exercise of authority and guidance, and control is felt as feedback about the effects of the action taken because thinking is required at all levels. (See figure.) This feedback allows the commander to adapt to changing circumstances and to command subsequent action. Cooperation is required in decentralized control. Subordinates work together laterally and from the bottom up to accomplish tasks that fulfill the commander's intent. Cooperation means we take the initiative to help those around us accomplish our shared mis- sion.[4]

Cooperation is essential to Marine Corps tactics. The flight leader and wingman work on the basis of cooperation. These pilots cooperate with the infantry they support. Two infantry units, fighting side by side, cooperate. A mobile combat service

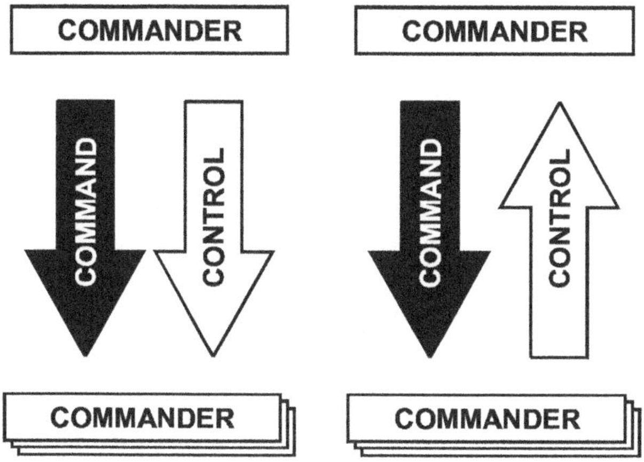


**A typical view of command and control—command and control seen as unidirectional.**

**Command and control viewed as reciprocal influence—command as initiation of action and control as feedback.**

**Two views of the relationship between command and control.**

support detachment and the mechanized force it supports cooperate. We all work together far more effectively when we communicate laterally than when we communicate only through higher headquarters and respond only to centralized direction. As an ancillary benefit, we relieve our overloaded communications networks.

The history of tactics is filled with examples where cooperation made the difference—and control could not. One such example occurred during an Iraqi counterattack in Operation Desert Storm. Black smoke from burning oil wells turned the day into night. A UH-1N Huey pilot used his night vision equipment to lead flights of AH-1W Cobras through near-zero visibility to attack Iraqi armored vehicles. The specially equipped Huey designated targets so that the Cobras could engage them at near pointblank range with antiarmor Hellfire missiles. For nearly 10 hours, the Huey pilot led flight after flight into the pitched battle, earning the Navy Cross for heroism.[5]

The pilots worked together to destroy targets the Huey could not engage and the Cobra could not see. This example shows what cooperation can accomplish.

## Discipline

Cooperation can harmonize efforts and get everyone to work together without the centralized control that undermines initiative. However, it raises a more fundamental question: How do we prepare people to cooperate when the going gets tough?

The answer is discipline. "There is only one kind of discipline—PERFECT DISCIPLINE. If you do not enforce and maintain discipline, you are potential murderers."[6] In the face of adversity and difficulty, discipline enables individuals to pursue what is best for those around them, their unit, and the Marine Corps. Individuals and units might have the desire, but without discipline they will be unable to accomplish the most difficult tasks in combat—operating faster than the enemy, gaining advantage, generating decisive force, and achieving decisive results.

In combat, instant obedience to orders is crucial. Orders may not be popular, but there comes a point where they must be carried out without question. Discipline is a result of training. In training for war, discipline should be firm, but fair. The Marine Corps is known as a highly disciplined fighting force. Discipline is one of the strengths that make Marines equally effective assaulting a beach, conducting a noncom-batant evacuation operation, fighting a fire, or guarding our embassies. Nonetheless, discipline is founded not only on obedience but also on a sense of duty.

The discipline needed for cooperation comes from two sources: imposed discipline and self-discipline. The first source, imposed discipline, is more often associated with the term "military discipline." Imposed discipline, typified by the Prussian approach, is characterized by instant obedience to orders. External in nature, it ensures compliance with established procedures, rules, or guidance and direction from above. It is a means to achieve efficiency in accomplishment of routine duties or drills. In its most extreme form, it can be rigid, paralyzing, and destructive of initiative. Imposed discipline also may make units vulnerable to the effects of chaos and uncertainty and unable to cooperate with one another.

Self-discipline is an internal force that morally obligates all Marines to do what they know is right—in this case, to cooperate with every other Marine in the pursuit of a common goal. The obligation is internal in each individual; it is something he or she feels powerfully about. Coupled with a sense of camaraderie and esprit de corps, it pulls from within and causes Marines to do everything they can for fellow Marines. At the unit level, this force can be felt as morale. "No system of tactics can lead to victory when the morale of an army is bad."[7]

Self-discipline can be seen in successful athletic teams as well as military units. Team players instinctively back up their teammates. In baseball, the outfielders cover each other on fly-balls. In hockey, rarely does only one player rush the goal. In football, offensive linemen do not stand by idly on a pass play if no defensive player faces them. They block the first defender they can find. Members of squads and fire teams also work together as teams to accomplish tasks and take care of each other. This cooperation among teammates cannot be enforced by a coach or leader. It depends upon the self-discipline of the individuals.

Marine discipline is the self-discipline of a successful team, not just the imposed discipline of the army of Frederick the Great. For Marines, military discipline means accepting personal responsibility. Self-discipline will not allow us to shirk responsibility or blame others. A discipline failure—often a failure to act—is a personal failure.

Our form of discipline is also *absolute*. There is no time off. Someone else may be in charge, but that does not absolve us from the responsibility to do everything we can to achieve the common goal. It does not reduce our responsibility to cooperate with fellow Marines in our unit and beyond.

This discipline is a mindset, *a way of thinking and behaving*. It runs through everything that we do. It is as much a part of garrison life as of combat. We also carry this sense of personal responsibility and duty to contribute into our private lives. We see it whenever off-duty Marines take the initiative to help at the scene of an accident, act as leaders in their communities, or in other ways do more than their share. They do so because of something inward, not because they are compelled by control. That something is self-discipline, and it is not limited to one aspect of life.

## Conclusion

Modern tactics relies on cooperation. Cooperation, in turn, depends on discipline. Discipline consists of both imposed discipline and self-discipline. As leaders of Marines, we must create a climate in which self-discipline and a high level of initiative can flourish within the boundaries of military discipline. This climate depends on us. Words are easy; anyone can give an occasional pep talk on the merits of self-discipline. Marines judge actions, not words, and respond positively to leadership by example. If the leader creates a climate where perfect discipline is expected and demonstrated, cooperation will follow.

# Chapter 7

# Exploiting Success and Finishing

*"Do not delay in the attack. When the foe has been split off and cut down, pursue him immediately and give him no time to assemble or form up . . . spare nothing. Without regard for difficulties, pursue the enemy day and night until he has been annihilated."*[1]

—Field Marshal Prince Aleksandr V. Suvorov

*"Pursue the last man to the Adda and throw the remains into the river."*[2]

—Field Marshal Prince Aleksandr V. Suvorov

*"When we have incurred the risk of a battle, we should know how to profit by the victory, and not merely content ourselves, according to custom, with possession of the field."*[3]

—Maurice de Saxe

It is not enough merely to gain advantage. The enemy will not surrender simply because he is placed at a disadvantage. The successful leader exploits any advantage aggressively and ruthlessly not once but repeatedly until the opportunity arises for a finishing stroke. We must always be on the lookout for such opportunities—whether we create them ourselves or they arise in the flow of action—and when we perceive an opportunity to be decisive, we must seize it.

The application of Marine Corps tactics does not mean that we expect to win effortlessly or bloodlessly or that we expect the enemy to collapse just because we outmaneuver him. It means we look for and make the most of every advantage and apply the decisive stroke when the opportunity presents itself.

## BUILDING ON ADVANTAGE

Once we have gained an advantage, we exploit it. We use it to create new opportunities. We then exploit those opportunities to create others, shaping the flow of action to our advantage. The advantages do not necessarily have to be large; even small favoring circumstances exploited repeatedly and aggressively can quickly multiply into decisive advantages. Like the chess grandmaster, we must think ahead to our next move and the one beyond it: *How am I going to use this advantage to create another one?* For example, in an attack by penetration, once we have created one advantage by punching through the enemy's position and getting into his rear, we create another by pouring forces through the gap, generating the "expanding torrent" that Liddell Hart wrote about.[4]

Rommel recounts how exploiting each advantage in the battle for Kuk in the Carpathian mountains during World War I led to another opportunity. As his detachment exploited each situation and moved farther behind the enemy lines, it generated more surprise and advantage. During this action, Rommel's detachment captured thousands of enemy soldiers with very little fighting, due largely to his unwillingness to lose momentum. One success led directly to another opportunity, which he immediately seized.[5]

After the battle for Tarawa in November 1943, Major Henry Crowe, Commanding Officer of 2d Battalion, 8th Marines, was asked why he thought the Japanese had been defeated so quickly once the Marines were established ashore. He remarked that it was due to the constant pressure of naval gunfire, bombs, and mortars. The Marines used their advantage in supporting arms to create opportunities for success.[6]

## Consolidation, Exploitation, and Pursuit

Once we have created leverage, how do we take advantage of it? A decisive result or victory rarely stems from the initial action, no matter how successful. More often, victories are the result of aggressively exploiting some relative advantage until one becomes decisive and the action turns into a rout. Casualty rates historically tend to remain relatively constant and often fairly even until one side or the other tries to flee. Only then do significantly asymmetrical casualty rates commonly occur. This exploitation of the enemy's bad situation can yield surprisingly great results.

We can take several specific types of actions to exploit opportunities we have created or discovered. The first way we can exploit success is by *consolidation*—as when we consolidate our forces after seizing a position we intend to hold against the enemy.[7] Here our aims are limited to protecting what we have already gained. We must realize that by consolidating, rather than continuing to force the issue, we may be surrendering the initiative. There may be any number of reasons for choosing this course. Perhaps we lack the strength to continue to advance. Our new gain may be of critical importance, and the risk of losing it outweighs the advantages of any further gains. Perhaps the new gain by itself grants a significant advantage. For instance, a position that provides excellent fires or threatens the enemy's lines of communications may put the enemy in an untenable position. Perhaps the new gain compels the enemy to meet us on our terms—for example, we seize a critical piece of terrain with strong defensive qualities, forcing the enemy to attack on unfavorable terms.

The second way to pursue an advantage is through *exploitation*, an offensive tactic that is designed to disorganize the enemy in depth.[8] Exploitation usually follows a successful attack that has created or exposed some enemy vulnerability. For example, an attack that has torn a gap in enemy defenses allows us to attack vital enemy rear areas. The object of exploitation is not to destroy the combat forces directly opposing us, even though they may be weakened. Instead, the object is to disrupt the entire enemy system by attacking important activities and functions.

For example, during Operation Desert Storm in 1991, the Army's Tiger Brigade was employed by the 2d Marine Division as an exploitation force during the division's final attack. The brigade had the advantage over the Iraqis in speed, firepower, and night combat capabilities. With these advantages the Tiger Brigade sliced deep into the rear of the Iraqi III Corps and sealed off the vital highway intersections north of Al Jahra. The result was a total disruption of the Iraqi organized defense.[9]

The third way to exploit advantage is through *pursuit*. A pursuit is an offensive tactic designed to catch or cut off a hostile force that has lost cohesion and is attempting to escape in order to destroy it.[10] If the intent is to bring about the final destruction or capture of the enemy's forces, then pursuit should be pushed with the utmost vigor. It is here that operations turn into routs, and overwhelming victories often occur.

General Grant's pursuit of General Lee's Confederate Army of Northern Virginia from Petersburg to Appomattox in April

1865 is a classic example of a pursuit. Here Grant pushed his forces to their limits in order to prevent Lee's escape. This ultimately led to the capture and surrender of Lee's forces.[11]

The Confederate Army's Lieutenant General Thomas J. "Stonewall" Jackson summed up pursuit when he said, "Strike the enemy and overcome him, never give up the pursuit as long as your men have strength to follow; for an enemy routed, if hotly pursued, becomes panic-stricken, and can be destroyed by half their number."[12]

## Finishing the Enemy

Ultimately, we want to cultivate opportunities into a decisive advantage. Once we do, we make the most of it. Marine Corps tactics calls for leaders who are "strong finishers." We must have a strong desire to "go for the jugular." We must be constantly trying to find or to create the opportunity to deliver the decisive blow. At the same time, we must not be premature in our actions. We must not make the decisive move before the conditions are right.

This ability to finish the enemy once and for all derives first from possessing an aggressive mentality. Second, it stems from an understanding of the commander's intent. Third, it stems from a keen situational awareness that helps us recognize opportunities when they present themselves and understand when the conditions are right for action.

## Use of the Reserve in Combat

The reserve is an important tool for exploiting success. The reserve is a part of the commander's combat power initially withheld from action in order to influence future action.[13] The reason to create and maintain a reserve is to provide flexibility to deal with the uncertainty, chance, and disorder of war. The reserve is thus a valuable tool for maintaining adaptability. In general, the more uncertain the situation, the larger should be the reserve. Napoleon once said that "War is composed of nothing but accidents, and . . . a general should never lose sight of everything to enable him to profit from [those] accidents."[14] These accidents take the form of opportunities and crises. The reserve is a key tactical tool for dealing with both.

The commander should have a purpose in mind for the reserve's employment and design it to fulfill that purpose. To truly exploit success may warrant assignment of the commander's best subordinate unit or a preponderance of combat power or mobility assets to the reserve. Those commanders who properly organize, task, and equip their reserves are usually the ones with the capability to finish the enemy when the opportunity arises.

Winston Churchill recognized the value of a reserve when he wrote: "It is in the use and withholding of their reserves that the great Commanders have generally excelled. After all, when once the last reserve has been thrown in, the Commander's part is played . . . . The event must be left to pluck and to the fighting troops."[15]

A strong reserve is also a way to retain the initiative. If an advance slows, the reserve can increase the momentum. If an advance picks up speed, the commitment of the reserve can create a rout. We may use the reserve to expand or exploit gaps or penetrations. We may commit the reserve to attack in a different direction, thus exploiting opportunities for success instead of reinforcing failure. Without a strong reserve, even the most promising opportunities can be lost.

A classic example of the use of the reserve is the battle for Tarawa. With the 2d and 8th Marine Regiments held up on the assault beaches, General Julian Smith decided to land the 6th Marine Regiment, the division reserve, to break the stalemate. The 1st Battalion, 6th Marines, which was task-or- ganized as part of the division reserve, landed on the western end of the island, passed through 3d Battalion, 2d Marines, and from the flank conducted a swift and violent assault of the Japanese fortifications across the island. Within 48 hours, the Japanese forces were annihilated and the island secured. General Smith's use of his reserve to exploit success and finish the enemy was the key to victory at Tarawa.[16] (See figure.)

Sometimes we must employ the reserve to deal with some crisis, rendering it temporarily unavailable for commitment elsewhere. In such instances, a reserve should be reconstituted as rapidly as possible. We should look for the opportunity to employ the reserve to reinforce success. However we may employ the reserve, we should always think of it as the tool for clinching the victory. In this respect, Marshal Foch wrote that "the reserve is a *club*, prepared, organized, reserved, carefully maintained with a view to carrying out the one act of battle

**Use of the Reserve:** Tarawa, 1943

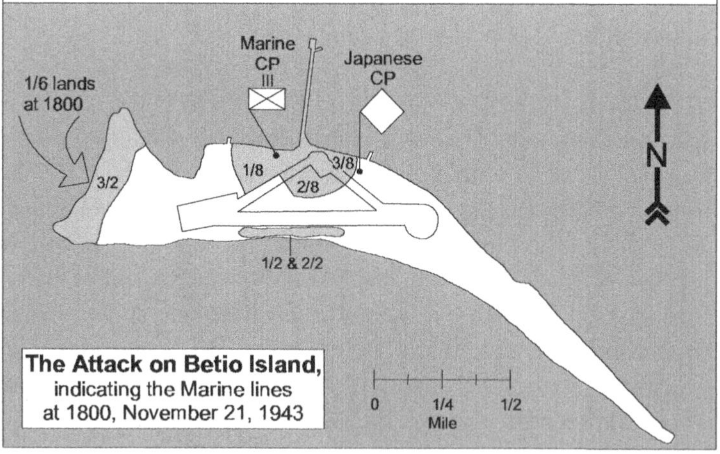


**The Attack on Betio Island,** indicating the Marine lines at 1800, November 21, 1943

from which a result is expected—the deci- sive attack."[17] It is generally through offensive action, even in the defense, that we achieve decisive results. Since the reserve represents our bid to achieve a favorable decision or to prevent an unfavorable one, it often becomes the main effort once committed and should be supported by all the other elements of the force.

Along with the tangible assets used as a reserve, the prudent commander must also be aware of, and plan for, the intangible factors that impact on combat power and its sustainment. Intangible factors include fatigue, leadership quality, proficiency, morale, teamwork, and equipment maintenance. We build reserves also by reserving aviation sortie rates or numbers, withholding unique or low-density munitions, or holding critical supplies such as fuel or petroleum, oils, and lubricants for a specific goal. We consider these intangible factors when creating and tasking the reserve, as we do in all assignments of tasks.

These concepts apply not only to units initially designated as the reserve but also to any unit, since any unit can be shifted or recommitted as the reserve. Thus a commander must always be mentally prepared to redesignate roles of units and to create and use reserves as the situation requires.[18]

## Conclusion

Most decisive victories do not result from the initial action, but from quickly and aggressively exploiting the opportuni- ties created by that action. We may find any number of ways to exploit tactical opportunity, but they all have the same object—to increase leverage until we have the final opportunity to decide the issue once and for all in our favor. A goal in Marine Corps tactics is not merely to gain advantage but to boldly and ruthlessly exploit that advantage to achieve final victory.

# Chapter 8

# Making It Happen

*"Nine-tenths of tactics are certain, and taught in books: but the irrational tenth is like the kingfisher flashing across the pool and that is the test of generals. It can only be ensured by instinct, sharpened by thought practicing the stroke so often that at the crisis it is as natural as a reflex."*[1]

—T. E. Lawrence

*"It cannot be too often repeated that in modern war, and especially in modern naval war, the chief factor in achieving triumph is what has been done in the way of thorough preparation and training before the beginning of war."*[2]

—Theodore Roosevelt

Reading and understanding the ideas in this publication are the initial steps on the road to tactical excellence. The primary way a Marine leader becomes an able tactician is through training and education, both of which are firmly rooted in doctrine. Doctrine establishes the philosophy and practical framework for how we fight. Education develops the understanding, creativity, military judgment, and the background essential for effective battlefield leadership. Training follows doctrine and develops the tactical and technical proficiency that underlies all successful military action. Individual and group exercises serve to integrate training and education, producing a whole that is greater than the sum of its parts. The lessons learned from training and operational experience then modify doctrine.

## DOCTRINE

Doctrine establishes the fundamental beliefs of the Marine Corps on the subject of war and how we practice our profession.[3] Doctrine establishes a particular way of thinking about war and our way of fighting, a philosophy for leading Marines in combat, a mandate for professionalism, and a common language. Doctrinal development benefits from our collective experience and distills its lessons to further education and training.

Our doctrine within the Marine Corps begins with the philosophy contained in MCDP 1, *Warfighting.* This philosophy underlies publications in the Marine Corps Warfighting Publications series that contain tactics, techniques, and procedures for specific functions. This body of thought helps form Marine tacticians through its implementation in education and training. (See figure.)

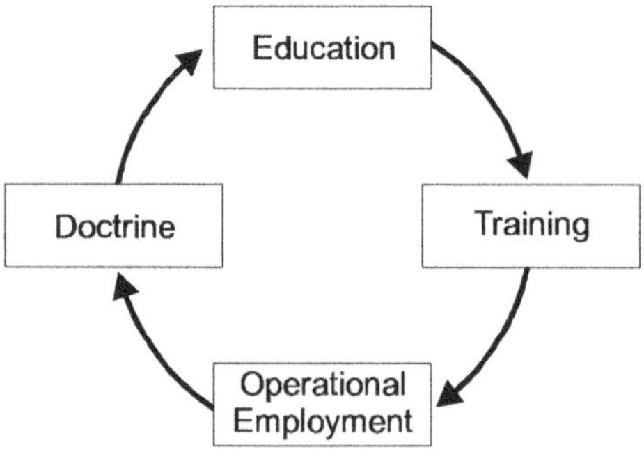


**The doctrinal development cycle.**

## EDUCATION

While combat provides the most instructive lessons on decisionmaking, tactical leaders cannot wait for war to begin their education. We must be competent in our profession before our skills are called upon. The lives of our Marines depend on it.

Our education in tactics must develop three qualities within all tactical leaders. The first quality is *creative ability.* Tactical leaders must be encouraged to devise and pursue unique approaches to military problems. No rules govern ingenuity. The line separating boldness from foolhardiness is drawn by the hand of practical experience. That said, an education in tactics must possess an element of rigor. Too often, tactical discussions lack an in-depth analysis of cause and effect. The tactically proficient leader must learn how to analyze solutions to tactical problems. Lacking such a rigorous analysis, the tactician will not learn from experience nor exercise creative ability.

The second quality is *military judgment*, which includes the skills for gaining situational awareness and acting decisively. The tactician must readily recognize the critical factors in any situation—enemy capabilities, weather, terrain characteristics, and the condition of our own forces, to mention just a few. Marine leaders must be able to cut to the heart of a situation by identifying its important elements, developing a sound plan, and making clear decisions. Our educational approach should emphasize the ability to understand the mission, issue a clear intent, and determine the main effort.

The third quality is *moral courage*. Moral courage is the ability to make and carry out the decision regardless of personal cost. It is different from—and rarer than—physical courage. The cost of physical courage may be injury or death, whereas the cost of moral courage may be the loss of friends, popularity, prestige, or career opportunities. The burden of conflicting responsibilities in combat—responsibility for the lives of subordinates, support for peers, loyalty to superiors, duty to the Nation—can be heavy. Our educational efforts should lead potential leaders to work through the proper resolution of such conflicts in peacetime. Leaders often need to make morally correct decisions in combat, but there will rarely be time for deep moral or ethical contemplation on the battlefield.

An effective leader willingly takes on the risks which come with military responsibilities. In that light, the greatest failing of a leader is a failure to lead. Two steadfast rules apply. First, in situations clearly requiring independent decisions, a leader has the solemn duty to make them. Whether the subsequent action succeeds or fails, the leader has made an honorable effort. The broad exercise of initiative by all Marines will likely carry the battle in spite of individual errors. Second, inaction and omission based on a failure of moral courage are much worse than any judgment error reflecting a sincere effort to act. Errors resulting from such moral failings lead not only to tactical setbacks but to the breakdown of faith in the chain of command. Proper training, education, and concerned leadership are the keys to instilling the qualities of creative ability, military judgment, and moral courage in the minds of all Marines.

## Training

Good tactics depend upon sound technical skills. These are the techniques and procedures which enable us to move, shoot, and communicate. We achieve technical competence through training. We build skills through repetition. Training also instills confidence in weapons and equipment. It develops the specialized skills essential to functioning in combat.

One of the ultimate aims of training is speed. Essential to speed is the requirement for accuracy. Speed without accuracy may be counterproductive and causes more damage than inaction. Whether Marines compute firing data, practice rifle marksmanship or weapons gunnery, rearm and refuel aircraft, repair vehicles, stock or transport supplies, or communicate information, the speed and accuracy of their actions determine the tempo of the overall force. Training develops the proficiency which enables this effective combination of speed and accuracy.

Small-unit training should focus on proficiency in such techniques and procedures as immediate-action drills, battle drills, and unit standing operating procedures. Practicing to reach technical proficiency applies to all types of units, whether a section of aircraft executing air combat maneuvers, a maintenance contact team repairing a vehicle under fire, an artillery gun team conducting displacement drills, or a rifle squad conducting an in-stride breach of an obstacle. We develop and refine these measures so that units gain and maintain the speed and accuracy essential for success in battle.

Staffs, like units and individual leaders, must train to increase speed and accuracy. Staffs increase speed by accomplishing three things: first, by obtaining and organizing information to help the commander and themselves understand the situation; second, by understanding the commander's decision and coordinating efforts to focus combat power to achieve the commander's goal; and third, by monitoring events, maintaining situational awareness, and anticipating and adapting to changes. As staffs train, they increase accuracy by becoming more proficient both in their respective areas and in functioning as a team.

Field Marshal Erwin Rommel knew the value of speed and accuracy for his staff when he wrote:

> A commander must accustom his staff to a high tempo from the outset, and continuously keep them up to it. If he once allows himself to be satisfied with norms, or anything less than an all-out effort, he gives up the race from the starting post, and will sooner or later be taught a bitter lesson.[4]

The speed and efficiency of a unit depend not only on the technical proficiency of its individual members but also in large part upon its cohesiveness. Such cohesion requires both personnel stability and solid leadership.

Training should also prepare Marines for the uniquely physical nature of combat. Living and caring for themselves in a spartan environment, confronting the natural elements, and experiencing the discomfort of being hungry, thirsty, and tired are as essential in preparing for combat duty as any skills training. The point is not to train individuals on how to be miserable, but rather on how to be effective when miserable or exhausted.

Likewise, training should enable us to take appropriate action in any environment and at any time. This readiness includes operating during inclement weather and periods of limited visibility. We must make terrain, weather, and darkness our allies if we are to gain advantage and deliver decisive force at a time and place of our choosing. We can neither anticipate nor appreciate the inherent friction that these natural factors produce unless we experience them.[5]

## TRAINING AND EDUCATIONAL METHODS

There is no single "best" approach to developing tactical proficiency. However, any approach should be adaptable to all echelons and to all grades. The environment should be one that is challenging and conducive to creative thinking. Like all preparation for war, training should reflect the rigors of that environment. The following examples may provide some tools for developing tactical proficiency in Marines.

**Professional Reading and Historical Study**

Because of the relative infrequency of actual combat experiences in most military leaders' careers, Marines must seek to expand their understanding through other, less direct means. The study of military history is critical to developing judgment and insight. It enables us to see how successful commanders have thought through—and fought through—the situations they faced. Not many people can do it instinctively—few possess the rare native ability to think militarily. Even those few can enhance their abilities through study and practice.

Historical studies provide the most readily available source of indirect experience in our profession. These studies describe the leadership considerations, the horrors of war, the sacrifices endured, the commitment involved, the resources required, and much more. These studies include biographies and autobiographies of military figures, books on specific battles, wars, and military institutions, unit histories, after-action reports, films, and documentaries. Group discussions help to expand the insights into leadership and battle that we have gained through individual study.

Professional readings and study are not solely the responsibility of military schools. Individuals cannot afford to wait for attendance at a military school to begin a course of self-directed study. Military professionalism demands that individuals and units find time to increase their professional knowledge through professional reading, professional military education classes, and individual study.

**Tactical Exercises**

Tactical success evolves from the synthesis of training and education—*the creative application of technical skills based on sound judgment.* Exercises enable leaders to practice decisionmaking and individuals, staffs, and units to practice and perfect collective skills. Exercises also serve to test and improve tactics, techniques, and procedures, immediate actions, battle drills, and combat standing operating procedures.

An exercise should serve as a unit's internal assessment of the quality of its training and education, not as grading criteria for higher commands. The conclusions should aim to note shortfalls so as to address them through future instruction and not to penalize poor performance. A unit will never be fully trained. There will always be room for improvement.

Exercises also test the ability of units to sustain tempo for an extended period of time. Since victory is rarely the product of single actions, the ability to operate and sustain combat effectiveness over time is important. Knowing when hostilities will cease is a convenience denied the combat Marine. Equipment must be maintained, and people must be sustained with adequate rest, nourishment, and hygiene until they accomplish their mission.

Tactical exercises can range from field exercises to command post exercises to tactical exercises without troops. Field exercises, conducted by units of any size, involve all unit personnel working together to learn, test, and refine their collective battlefield tasks. Such exercises can be general in nature, or they can be detailed rehearsals for specific upcoming missions.

Command post exercises are largely limited to commanders and their staffs. Their purpose is to familiarize staffs with their commanders' personal preferences and operating styles as well as to exercise staff techniques and procedures and to review particular contingency plans.

Tactical exercises without troops provide tactical leaders opportunities to exercise judgment while permitting other unit elements to conduct training and education of their own. There are two approaches to conducting them.

The first method provides a leader an opportunity to evaluate a subordinate's ability to perform in a given scenario. This method places students in an area of operations and provides a situation upon which to plan and execute a task—for example, "Establish a reverse slope defense." The aim here is to exercise tactical proficiency in the siting of weapons and the use of terrain.

The second method also places students in an area of operations and provides a situation but gives them a mission order—for example, "Prevent enemy movement north of Route 348." The aim here is to exercise judgment. After walking the ground, the students must first decide whether to defend or attack, supporting their conclusions with reasoning. The reasoning is then discussed and criticized. This approach encourages students to demonstrate ingenuity and initiative. They have free rein to employ their resources as they see fit to achieve the desired results.[6]

### Wargaming

Wargames can be a valuable tool for understanding the many factors that influence a leader's decisions. Morale, enemy and friendly situations, the higher commander's intentions, firepower, mobility, and terrain are only a few of the decision factors included in the play of wargames. In all these simulations, from the sand table to a commercial board game to a computerized simulation, routine should be avoided. The less familiar the environment, the more creativity the student must display.

Sand table exercises, tactical decision games, and map exercises present students with a general situation, mission orders, and a minimum of information on enemy and friendly forces. Sand table exercises are especially suited to novice tacticians. They present the terrain in three-dimensional array, whereas a map requires interpretation. Both map and sand table exercises enable students to conceptualize the battle, deliver their decisions, and issue orders to subordinates. Afterwards, students discuss their decisions and are critiqued. The discussion should focus on making a decision in the absence of perfect information or complete intelligence.

### Terrain Walks

Terrain walks introduce the realities of terrain, vegetation, and weather. Terrain walks can be conducted in at least two ways.

The first method provides students with an area of operations, a general situation (usually depicted on a map), and a mission. As in sand table and map exercises, students describe their view of the battle. Choosing one plan, the group then begins to walk the terrain according to the plan. The group will then encounter unanticipated terrain and obstacles, while the instructors introduce enemy actions into the play of the problem. In this way, students must contend with the disparity between actual terrain and vegetation and maps as well as the chaos and uncertainty generated by enemy actions that invariably occur in real-world operations.

The second method involves the firsthand study of historic battlefields. We gain a special vantage on battle by walking the ground and seeing the battlefield from the perspective of both commanders. We gain a new appreciation for an historical commander's blunders. Often such blunders seem incomprehensible—until we see the ground. Only then can we realistically consider alternative courses of action that the commander might have pursued.[7]

### Competition

Exercises should provide realism. The means to achieve tactical realism are competitive free-play or force-on-force exercises. Whenever possible, unit training should be conducted in a free-play scenario. This approach can be used by all leaders to develop their subordinates. It affords both leaders and unit members the opportunity to apply their skills and knowledge against an active threat.

Free-play exercises are adaptable to all tactical scenarios and beneficial to all echelons. Whether it is fire teams scouting against fire teams, sections of aircraft dueling in the sky, or companies, battalions, squadrons, and Marine air-ground task forces operating against one another, both leaders and individual Marines benefit. Leaders form and execute their decisions against an opposing force as individual Marines employ their skills against an active enemy. Through free-play exercises, Marines learn to fight as an organization and to deal with a realistically challenging foe.[8]

### Critiques

A key attribute of decisionmakers is their ability to reach decisions with *clear reasoning*. Critiques elicit this reasoning process. Any tactical decision game or tactical exercise should culminate with a critique.

The standard approach for conducting critiques should promote initiative. Since every tactical situation is unique and since no training situation can encompass more than a small fraction of the peculiarities of a real tactical situation, there can be no ideal or school solution. Critiques should focus on the students' rationale for doing what they did. What factors did a student consider, or not consider, in making an estimate of the situation? Were the decisions the student made consistent with this estimate? Were the actions ordered tactically sound? Did they have a reasonable chance of achieving success? How well were the orders communicated to subordinates? These questions should form the basis for critiques. The purpose is to broaden a leader's analytical powers, experience level, and base of knowledge, thereby increasing the student's creative ability to devise sound, innovative solutions to difficult problems.

Critiques should be open-minded and understanding, rather than rigid and harsh. Mistakes are essential to the learning process and should always be cast in a positive light. The focus should not be on whether a leader did well or poorly, but rather on the progress achieved in overall development. We must aim to provide the best climate to grow leaders. Dam- aging a leader's self-esteem, especially in public, therefore should be strictly avoided. A leader's self-confidence is the wellspring from which flows the willingness to assume responsibility and exercise initiative.[9]

## Conclusion

In this publication, we have explored themes that help us to understand the fundamentals and to master the art and science of tactics. From the study of our warfighting philosophy, we have gained an appreciation for the requirement to be decisive in battle. To accomplish this, we must clearly visualize the battlespace through gained situational awareness, recognize patterns, and make decisions intuitively. We have also discussed ways we can gain advantage over the enemy and force him to bend to our will. We also explored how to be faster in relation to the enemy, to adapt to changing conditions, to cooperate for success, to exploit success, and to finish the enemy. Finally, we discussed how we can begin to act on these ideas during our training for combat. The ideas presented in this publication have implications far beyond battlefield tactics and the doctrinal way we think about warfare. They also influence the way we organize—using task organization and flexible command and control relationships—and the way we equip ourselves for combat.

Waging war in maneuver warfare style demands a professional body of officers and Marines schooled in its science and art. When asked why the Marines were so successful in Operation Desert Storm, General Boomer replied:

> The thing that made the big difference on the battlefield is that we had thousands and thousands of individual Marines constantly taking the initiative. The young lance corporal

> would take a look, see something 75 or 100 meters out in front that needed to be done, and go out and do it without being told. As I read through [the] award citations from Desert Shield and Desert Storm, this theme reappears, time and time again. That aggressive spirit comes from being well-trained, and confident in your professional knowledge.[10]

Everything we do in peacetime should prepare us for combat. Our preparation for combat depends upon training and education that develop the action and thought essential to battle.

## Understanding Tactics

1. Statement by Gen A. M. Gray, former Commandant of the Marine Corps, during a ceremony commemorating the anniversary of the groundbreaking for the Marine Corps Research Center, June 20, 1997.

2. Sir William Slim, *Defeat into Victory* (London: Cassell and Co. Ltd., 1956) pp. 550-551.

3. MCDP 1, *Warfighting* (June 1997) p. 30. MCDP 1's definition differs from that given in Joint Pub 1-02, *Department of Defense Dictionary of Military and Associated Terms*: "**tactics**—1. The employment of units in combat. 2. The ordered arrangement and maneuver of units in relation to each other and/or to the enemy in order to use their full potentialities."

4. Ibid., p. 3.

5. **Combat power**: "The total means of destructive and/or disruptive force which a military unit/formation, can apply against the opponent at a given time." (Joint Pub 1-02)

6. LtCol G. I. Wilson, "The Gulf War, Maneuver Warfare, and the Operational Art," *Marine Corps Gazette* (June 1991) pp. 23–24.

7. This example was taken from Joseph H. Alexander, *Utmost Savagery; The Three Days of Tarawa* (Annapolis, MD: Naval Institute Press, 1995).

8. Bill D. Ross, *Iwo Jima: Legacy of Valor* (NY: Vanguard Press, 1985) pp. 79–80.

9. Carl von Clausewitz, *On War*, trans. and eds. Michael Howard and Peter Paret (Princeton, NJ: Princeton University Press, 1984) p. 119.

10. *Medal of Honor Recipients 1863–1973* (Washington, D.C.: U.S. Government Printing Office, 1973) p. 492.

11. Clausewitz, p. 85.

12. *Infantry in Battle* (Washington, D.C.: The Infantry Journal, Incorporated, 1939) p.1.

## Achieving a Decision

1. *Infantry in Battle*, p. 1.

2. FMFM 1, *Warfighting* (March 1989) p. 61.

3. Martin Blumenson, *Anzio: The Gamble That Failed* (Philadelphia, PA: J. B. Lippincott Company, 1963).

4. Shelby Foote, *The Civil War: A Narrative* (NY: Random House, 1963) pp. 467–468.

5. Maj Charles D. Melson, Evelyn A. Englander, Capt David A. Dawson, comps., *U.S. Marines in the Persian Gulf, 1990–1991: Anthology and Annotated Bibliography* (Washington, D.C.:

Headquarters, U.S. Marine Corps, History and Museums Division, 1992) p. 181. Also see pages 173–182 of the same publication and LtCol Charles H. Cureton, *U.S. Marines in the Persian Gulf, 1990–1991: With the 1st Marine Division in Desert Shield and Desert Storm* (Washington, D.C.: Headquarters, U.S. Marine Corps, History and Museums Division, 1993) pp. 26–27.

## Gaining Advantage

1. Robert Debs Heinl, Jr., Col, USMC, Retired, *Dictionary of Military and Naval Quotations* (Annapolis, MD: United States Naval Institute, 1966) p. 321.

2. FMFM 1, p. 57.

3. **Combined arms:** "The tactics, techniques, and procedures employed by a force to integrate firepower and mobility to produce a desired effect upon the enemy." FMFRP 0-14, *Marine Corps Supplement to the DOD Dictionary of Military and Associated Terms* (January 1994).

4. Martin Blumenson, *The Patton Papers*, vol. 2 (Boston, MA: Houghton Mifflin Company, 1974) pp. 39–40.

5. Andrew Geer, *The New Breed* (NY: Harper & Brothers, 1952) p. 339.

6. Ibid., pp. 365–366.

7. Joe Douglas Dodd, "Night Attack on Kunishi Ridge," *Marine Corps Gazette* (April 1985) p. 43.

8. Ibid., pp. 42–44.

9. Sun Tzu, *The Art of War*, trans. Samuel B. Griffith (NY: Oxford University Press, 1963) p. 91.

10. Ibid., p. 106.

11. Maj Robert R. Parker, Jr., "Deception: The Missing Tool," *Marine Corps Gazette* (May 1992) p. 97.

12. Sun Tzu, p. 98.

13. *The Marines in Vietnam, 1954-1973: An Anthology and Annotated Bibliography* (Washington, D.C.: Headquarters, U.S. Marine Corps, History and Museums Division, 1985) pp. 173–181.

14. FMFM 6-7, *Scouting and Patrolling for Infantry Units* (January 1989) p. 2-1.

15. Sun Tzu, p. 93.

## Being Faster

1. Heinl, p. 220.

2. John A. English, *On Infantry* (NY: Praeger, 1984) p. 223.

3. Jeter A. Isley and Philip A. Crowl, *The U.S. Marines and Amphibious War: Its Theory, and Its Practice in the Pacific* (Princeton, NJ: Princeton University Press, 1951) p. 338.

4. Peter G. Tsouras, *Warrior's Words: A Quotation Book: From Sesostris III to Schwarzkopf, 1871 B.C. to A.D. 1991* (London: Cassell Arms and Armour, 1992), p. 434.

5. Heinl, p. 63.

6. "Command," *Time* (January 25, 1943) p. 61.

7. Col Ray Smith, USMC, telephone interview by Capt S. R. Shoemaker, USMC, 12 March 1991, Washington, D.C..

8. Gen Hermann Balck, interview by William S. Lind, 6 June 1980, Washington, D.C..

9. William S. Lind, *Maneuver Warfare Handbook* (Boulder, CO: Westview Press, 1985) pp. 5–6.

10. Gen George S. Patton, Jr., *War As I Knew It* (NY: Bantam Books, Inc., 1979) p. 323.

11. Ibid., pp. 330–331.

12. English, p. 217.

13. Capt A. T. Mahan, USN, *The Life of Nelson: The Embodiment of the Sea Power of Great Britain* (Boston: Little, Brown, and Co., 1899) p. 730.

14. FMFRP 12-110, *Fighting on Guadalcanal* (September 1991) p. 33.

15. Patton, p. 376.

## Adapting

1. Tsouras, p. 21.

2. Ibid., p. 21.

3. Keith William Nolan, *Battle for Hue* (Novato, CA: Presidio Press, 1983) pp. 51–52.

4. **Branch**: "A contingency plan or course of action (an option built into the basic plan or course of action) for changing the mission, disposition, orientation, or direction of movement of the force to aid success of the operation based on anticipated events, opportunities, or disruptions caused by enemy actions and reactions as determined during the wargaming process." MCRP 5-2A, *Operational Terms and Graphics* (June 1997).

5. **Sequel**: "Major operations that follow the current major operation. Plans for these are based on the possible outcomes (victory, stalemate, or defeat) associated with the current operation." (MCRP 5-2A).

6. See MCDP 5, *Planning*, for a more complete discussion of modular plans.

7. Cureton, p. 77.

8. Ibid, p. 77.

## Cooperating

1. NAVMC 7386, *Tactical Principles* (Quantico, VA: Marine Corps Schools, 1955) p. 7–8.

2. MCDP 6, *Command and Control* (October 1996) p. 48.

3. Ibid., p. 43.

4. Ibid., pp. 39–41.

5. Melson, Englander, and Dawson, comps., p. 140.

6. Patton, p. 376.

7. Heinl, p. 196.

## Exploiting Success and Finishing

1. Tsouras, p. 349.

2. Ibid., p. 349.

3. Heinl, p. 109.

4. Capt B. H. Liddell Hart, "The 'Man-in-the-Dark' Theory of Infantry Tactics and the 'Expanding Torrent' System of Attack," *Journal of the R.U.S.I.* (February 1921) p. 13.

5. Erwin Rommel, *Attacks* (Vienna, VA: Athena Press, 1979) pp. 235–250.

6. FMFRP 12-90, *Second Marine Division Report on Gilbert Islands—Tarawa Operation* (September 1991) p. 51.

7. **Consolidation of position**: "Organizing and strengthening a newly captured position so that it can be used against the enemy." (Joint Pub 1-02)

8. **Exploitation**: "An offensive operation that usually follows a successful attack and is designed to disorganize the enemy in depth." (Joint Pub 1-02)

9. Col Charles J. Quilter, II, *U.S. Marine Corps in the Persian Gulf, 1990–1991: With the I Marine Expeditionary Force in Desert Shield and Desert Storm* (Washington, D.C.: Headquarters, U.S. Marine Corps, History and Museums Division, 1993) p. 99.

10. **Pursuit**: "An offensive operation designed to catch or cut off a hostile force attempting to escape, with the aim of destroying it." (Joint Pub 1-02)

11. Bruce Catton, *This Hallowed Ground: The Story of the Union Side of the Civil War* (Garden City, NY: Doubleday & Company, Inc., 1956) p. 384.

12. Heinl, p. 259.

13. **Reserve**: "Portion of a body of troops which is kept to the rear, or withheld from action at the beginning of an engagement, available for a decisive moment." **Tactical reserve**: "A part of a force, held under the control of the commander as a maneuvering force to influence future action." (Joint Pub 1-02)

14. Brig Gen Thomas R. Phillips, U.S. Army, ed., "Military Maxims of Napoleon," in *Roots of Strategy: A Collection of Military Classics* (Harrisburg, PA: Military Service Publishing Co., 1940) p. 436.

15. Heinl, p. 275.

16. This example was taken from Martin Russ, *Line of Departure: TARAWA* (Garden City, NY: Doubleday & Company, Inc., 1975).

17. Heinl, p. 274.

18. Much of the material in this section is based on Capt John F. Schmitt's article, "The Use of the Reserve in Combat," *Marine Corps Gazette* (March 1990) pp. 63–69.

## Making It Happen

1. T. E. Lawrence, "The Science of Guerrilla Warfare," introduction to "Guerrilla Warfare," *Encyclopedia Britannica*, 13th ed. (NY: Encyclopedia Britannica, 1926).

2. Heinl, p. 329.

3. **Doctrine**: "Fundamental principles by which the military forces or elements thereof guide their actions in support of national objectives. It is authoritative but requires judgment in application." (Joint Pub 1-02)

4. As attributed to Erwin Rommel by Heinl, p. 60.

5. For more detailed information on the establishment of unit training programs, see MCRP 3-0A, *Unit Training Management Guide* (November 1996).

6. For more detailed readings on the subject of designing and executing training exercises, see MCRP 3-0B, *How to Conduct Training* (November 1996).

7. William Glenn Robertson, *The Staff Ride* (Washington, D.C.: U.S. Army Center of Military History, 1987) provides an excellent description of the use of terrain walks. See also *Staff Ride Handbook* (Quantico, VA: Marine Corps University, 1996).

8. This section reflects the emphasis found in MCDP 1, *Warfighting,* regarding the requirement to simulate the "clash of opposing wills" found in combat by conducting free-play exercises.

9. The subject of how to conduct critiques and hold after-action reviews is covered in detail in both MCRP 3-0A and MCRP 3-0B.

10. Melson, Englander, and Dawson, comps., p. 94.

Made in the USA
Coppell, TX
10 November 2026

11835557R00090